AF344499

Embedded Systems

Series Editors

Nikil D. Dutt, Department of Computer Science, Donald Bren School
    of Information and Computer Sciences, University of California, Irvine,
    Zot Code 3435, Irvine, CA 92697-3435, USA
Peter Marwedel, Informatik 12, TU Dortmund, Otto-Hahn-Str. 16,
    44227 Dortmund, Germany
Grant Martin, Tensilica Inc., 3255-6 Scott Blvd., Santa Clara, CA 95054, USA

For other titles published in this series, go to
www.springer.com/series/8563

Paul Lokuciejewski · Peter Marwedel

Worst-Case Execution Time Aware Compilation Techniques for Real-Time Systems

 Springer

Paul Lokuciejewski
Suitbertusstr. 34
40223 Düsseldorf
Germany
Paul.Lokuciejewski@tu-dortmund.de

Peter Marwedel
Embedded Systems Group
TU Dortmund University
Otto-Hahn-Str. 16
44221 Dortmund
Germany
peter.marwedel@tu-dortmund.de

ISBN 978-90-481-9928-0   e-ISBN 978-90-481-9929-7
DOI 10.1007/978-90-481-9929-7
Springer Dordrecht Heidelberg London New York

Library of Congress Control Number: 2010936440

Cover design: VTEX, Vilnius

Printed on acid-free paper

Springer is part of Springer Science+Business Media (www.springer.com)

Acknowledgements

Many people have contributed in different ways to this work. We would like to thank Sascha Plazar, Jan C. Kleinsorge, Lars Wehmeyer, Timon Kelter, Daniel Cordes, and Florian Schmoll for the numerous technical discussions. Special thanks go to Heiko Falk for years of intense and inspiring cooperation, discussions and invaluable support. We would also like to thank the people from AbsInt Angewandte Information GmbH for the technical discussions and their software support.

This work has been partially funded by the European Community's ArtistDesign Network of Excellence and by the European Community's Seventh Framework Programme FP7/2007-2013 under grant agreement n° 216008.

Above all, we are deeply indebted to our families for their love and support helping us to write this book.

Dortmund   Paul Lokuciejewski
June 2010   Peter Marwedel

Contents

List of Figures

List of Tables

Chapter 1
Introduction

Contents

Computer systems are of vital importance to human life. In the previous decades, computers operated as large mainframes to assist people in their professional activities by processing and storing data. Following this era, personal computers have revolutionized the nineties by providing computation power to a vast portion of the world population. The trend of miniaturization has continued up to these days, leading to small and powerful systems that are often integrated into larger products.

While mainframes and desktop computers were stationary, the systems of the future change the perception of computer technology. They accompany the user permanently everywhere, either as small devices carried around or as part of systems encountered every day. Due to their disappearance by weaving into larger services, terms such as *pervasive computing* or *ubiquitous systems* [Wei99] have become popular, expressing the characteristics of this next-generation information technology.

The technical basis for this technology are cooperating *embedded systems*. The main characteristics of these systems are their integration into a larger product and information processing as their field of application [Mar10]. The rapidly increasing performance and the drop in the price of processors employed in embedded systems make them attractive to a large number of applications.

The following, incomplete list provides an overview of the major application domains for embedded systems as well as their properties:

Telecommunication: The most prominent example in this domain are mobile phones with a continuously growing market. Nowadays, these devices do not only provide functionality for communication but also features like a digital camera or internet surfing. Other examples are telecommunication switching centers for the coordination of different communication services.

P. Lokuciejewski, P. Marwedel, *Worst-Case Execution Time Aware Compilation Techniques for Real-Time Systems,* Embedded Systems, DOI 10.1007/978-90-481-9929-7_1, © Springer Science+Business Media B.V. 2011

Consumer Electronics: Modern devices have to cope with an increasing amount of video and audio data. In addition, companies can only survive on the market when products with new services and an improved quality are offered. Such high requirements can only be guaranteed when high-performance embedded systems are involved. Examples for consumer electronics are navigation systems, PDAs, television, or game consoles.

Automotive/Avionics: Traditionally, cars were considered as the typical domain of electrical and mechanical engineers. This view has dramatically changed in the last years. The current trend in the automotive and avionics domain goes towards the integration of many services on powerful electronic hardware platforms. Electronics and software represent a continuously increasing share in the added value of automotive products constituting up to 90% of all automotive innovations [FBH+06]. For example, modern cars like the BMW 7-series are equipped with more than 70 different processors executing more than 60 MB of software [Saa03]. In a similar fashion, modern aircrafts can not be flown without the assistance of computers.

All these application domains impose high requirements on modern embedded systems. Among economical requirements, such as low development costs and short time-to-market [Vah02], the devices must satisfy various non-functional requirements. The following are considered to be highly important objectives for current embedded systems:

- dependability
- temperature efficiency
- energy efficiency
- average-case performance
- worst-case performance

Embedded systems are becoming more and more pervasive, leading to an increasing reliance on their continual provision of correct services. Hence, these systems have to be highly dependable. For example, the system of a car should be fault-tolerant, i.e., continue operating correctly after a fault, safe and easy to maintain in a reasonable amount of time. In the context of reliability, temperature plays an important role. Thermal hot spots degrade reliability, but also have a negative impact on system performance, and increase cooling costs and leakage power. One way to reduce peak temperatures and the frequency of hot spots is the application of temperature-aware scheduling policies [HXV+05].

Energy efficiency plays a crucial role for many embedded systems, in particular in the domain of consumer electronics. Since a significant portion of these devices operates as portable systems on battery power, the battery lifetime decides about the success of the product. Devices that have to be frequently charged will be hardly accepted by the consumer. The energy optimization process can be considered as an interdisciplinary work between hardware and software designers. As demonstrated in different studies, slow main memory accounts for up to 70% of the system's power budget [KC02]. To diminish this effect, system designers place small, low-power memories close to the processor. A popular example for highly efficient memories are *scratchpads* which, however, require explicit support from the compiler for their utilization. Examples for energy-aware scratchpad allocation techniques are found in [WM06, VM07].

Simultaneously, the utilization of small memories has a positive effect on the system's average-case performance. The latter denotes the performance of a system that is run with a representative set of input data. Due to the increasing performance gap between large memories and processors [Mac02], the performance of the system is not governed by the processor but by the slow memory. This effect is known as the *memory wall problem* [WM95]. Exploiting small memories reduces the number of processor idle cycles resulting from the bottleneck to slow memories. Further hardware features found in modern processors to improve the average-case performance are caches, branch prediction, or speculative execution. These performance enhancing features automatically shorten the program execution for the average case but also introduce a variance in the execution time of operations. A well-known example is a cache-based system where the execution time of an operation may significantly vary depending on whether a cache hit or cache miss was encountered. Besides hardware optimization, system designers may improve the performance of software by the utilization of optimizing compilers. In the previous decades, a large portfolio of *average-case execution time* (*ACET*) optimizations has been developed [ASU86, PW86, Muc97]. With the increasing popularity of embedded systems, various compiler optimizations dedicated to these hardware platforms were presented in the last years [Leu00].

Embedded systems often interact with their physical environment. To express their strong link to physics, these systems are also referred to as *cyber-physical systems* [Lee07]. In a physical environment, temporal requirements have to be fulfilled. Primarily in safety-critical application domains, such as automotive or avionics, timing deadlines must be satisfied. Otherwise, physical damage or even loss of life could occur. For example, it is essential to know if an airbag in a car will fire fast enough to save lives. To satisfy stringent timing constraints, offline guarantees have to be derived, serving as a specification of the worst-case performance of the system. This performance is defined by the worst-case input data and a system state that leads to the maximal execution time of the system. If these offline guarantees are known, the system designer can ensure that timing constraints are met by either reducing the functionality of the original application, or by choosing a hardware platform where the executed tasks do not exceed their deadlines. Due to these critical timing constraints but also the high performance requirements imposed on modern embedded systems, their design represents a challenging multi-objective optimization problem.

In the following, challenges faced in the design process of time-critical embedded systems are discussed in more detail since the techniques developed in this book aim at an optimization of these devices.

1.1 Design of Embedded Real-Time Systems

Correct temporal behavior is a necessary property of systems called *real-time systems*. By definition, the correctness of a real-time system depends not only on

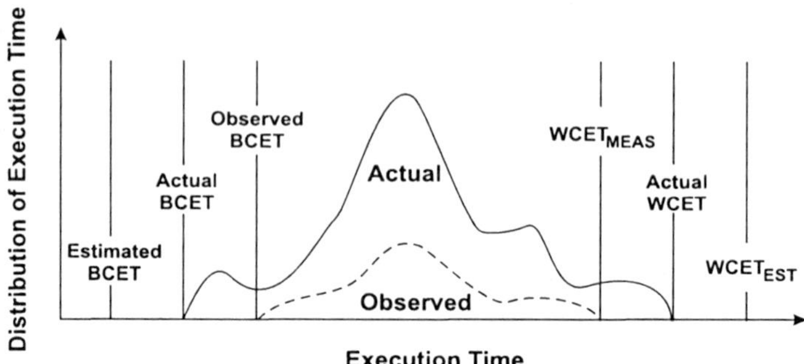

Fig. 1.1 Execution time distribution

the results of the computation, but also on the time at which the results are produced [Sta88]. Systems where the deadline must not be missed under any circumstances are termed *hard real-time systems*. An important parameter to reason about the timeliness is the *worst-case execution time (WCET)*. Figure 1.1 depicts the relevant timing properties of a real-time task which represents a unit of scheduling by an operating system. A task exhibits a certain variation of execution time depending on input data and different behaviors of the environment. The horizontal axis represents possible execution times while the vertical axis denotes their distribution. The WCET is the longest execution time that can ever occur. To find the maximal execution time it is, however, not feasible to exhaustively explore all possible execution times since software of realistic sizes that is run on modern processor architectures exhibits a too large state space. A simulation of the task with predefined input sets will possibly not observe the actual WCET. Therefore, these measurements deliver insufficient results (see $WCET_{MEAS}$) for the verification of a hard real-time system.

To reason about the WCET, formal methods are used instead. They analyze the system statically in order to derive sound properties of its temporal behavior. However, due to undecidability it is in general not possible to estimate the WCET, otherwise one could solve the halting problem. Thus, the WCET can only be computed for a restricted set of programs that meet certain constraints: the program under analysis is guaranteed to terminate and recursion depths as well as loop iteration counts must be explicitly bounded.

But even with these restrictions, it is impossible to compute the actual WCET. Modern processors are equipped with complex architectural features, such as superscalar pipelines or caches, which can often not be precisely analyzed by static methods. Any encountered uncertainties must be conservatively approximated and yield overapproximations. Furthermore, a sound approximation of the actual WCET demands an abstraction of possible inputs and initial states of the system. This mandatory abstraction introduces another source of imprecision. As a consequence, the determination of the actual WCET has to be relaxed to the derivation of an upper bound on the execution time of the task. These bounds represent the *estimated* WCET (cf. $WCET_{EST}$ in Fig. 1.1). The process of estimating the WCET is called *timing analysis*.

A reliable timing analysis must satisfy the following constraints:

- **Safeness**: $\text{WCET} \leq \text{WCET}_{\text{EST}}$
- **Tightness**: $\text{WCET}_{\text{EST}} - \text{WCET} \rightarrow 0$

Safeness is a mandatory property that must be met by any WCET analysis. Tightness serves as a metric for precision since tighter estimations reflect the actual WCET more precisely.

A notion related to the WCET is the *best-case execution time* (*BCET*) which represents the shortest execution time. In this work, the BCET will be not explicitly considered. However, the presented analyses and optimizations can be easily extended to cover this metric.

1.1.1 Industrial Practice for Meeting Timing Constraints

The goal of the design of embedded real-time systems is the development of an efficient system with a high worst-case performance, i.e., a system specified by a low WCET. Optimizing the worst-case performance in the development phase is crucial for the success of the product. The reduction of the WCET may lead to a safer system since timing deadlines which may be missed before the optimization are guaranteed to be satisfied. Moreover, an improved worst-case performance can significantly cut the product costs since cheaper hardware may be utilized for the optimized system while still satisfying timing constraints.

However, the current design process fails to deliver embedded real-time systems with a high worst-case performance. This is due to the fundamental problem of time-critical system design: the lack of timing in embedded software [Vah08]. The fact that no notion of time is available constitutes a common problem in the automotive and avionics application domain.

A prominent example from industry is AUTOSAR [Ric06], an initiative founded by leading car manufacturers to establish standards for automotive electrics/electronics (E/E) architectures. The initial missing specification of timing requirements in the meta-models of the project is considered as one of its major weak points [Ric06]. A similar situation can be observed in safety-critical avionics software running on *integrated modular avionics* (*IMA*) architectures [SLPH+05, GP07]. Researchers have recognized this crucial issue and appeal for a reinvention of computer science with regard to a clear notion of time [Lee05].

As a consequence of the missing timing-aware software development tools (like compilers), it is common practice in industrial environments to tune the worst-case performance by a *trial-and-error* based approach. This is illustrated in Fig. 1.2. The design process starts with a specification of the embedded systems software. Due to the complexity of embedded applications, systems are often modeled graphically using software engineering tools like ASCET [ETA10]. These tools generate code in a high-level language, predominantly ANSI C [Sch00], which is translated into machine code by a compiler.

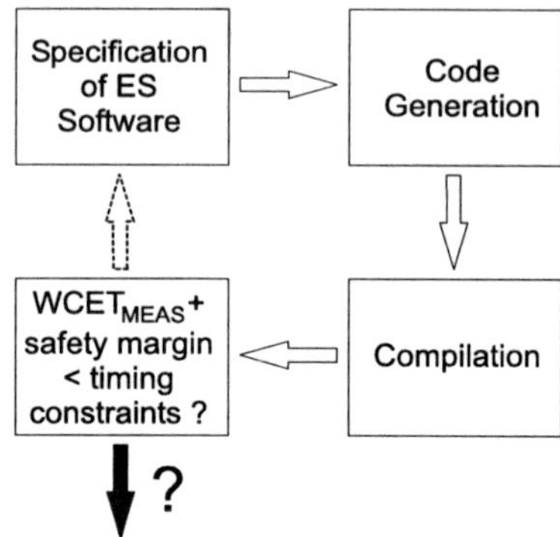

Fig. 1.2 Trial-and-error based design process

In the next step, the generated binary executable is utilized for WCET measurements. As it is known that measurements may yield an unsafe under-approximation of the actual WCET, system designers add a *safety margin* to the measured value. This accumulated value, which is considered as the system's WCET, is used for the verification of timing constraints. In case of violated constraints, the design has to be modified based on the developer's intuition. For this purpose either the software specification in the first step or the generated code in the second step (cf. Fig. 1.2) is manually modified. After another compilation phase, a repetitive verification of real-time constraints follows. This cyclic process is repeated until the resulting (measured) WCET promises to meet the system's timing specifications.

The manual modification of the software specification or the program code is challenging since the well-known principles of ACET optimizations are not applicable: optimizing the most frequently executed code may not succeed for real-time systems. Hence, the system designer is faced with the following two central issues:

1. **Subject of Optimization**: What should be modified in the software to improve the worst-case performance?
2. **Quantification:** Which impact does a modification of the software exhibit on the worst-case performance?

Since the answer to these questions is often not known, the iterative process of code generation and verification leads to high development times. As the number of required trials to meet timing constraints is hard to predict, the development time becomes non-deterministic. This situation does not conform to today's short time-to-market periods.

The design shown in Fig. 1.2 may also boost production costs since the developers are forced to utilize oversized and expensive hardware. This is due to two reasons. First, manual tuning of code usually yields poorly optimized software. Second, the safety margin, which is added to the measured WCET, may be chosen too large. As a result, system resources are wasted.

The above discussion emphasizes that the current trial-and-error based approach used in industry is not suitable for an efficient design of competitive embedded real-time systems. Besides, the common use of WCET measurements is highly unsafe since no guarantees can be deduced that the observed run times do not violate timing constraints.

The first step to improve this industrial design flow would be the substitution of WCET measurements by a static timing analyses. This way the employed timing information becomes more precise since a safety margin is not required any more. Consequently, cheaper hardware can be used for the development. Moreover, the considered WCET information gets safe, removing the danger of potentially violating timing constraints. However, this modification of the design flow does not help to avoid the inefficient trial-and-error based tuning since the need for manual software modifications remains. Hence, long development times and low-performance code persist.

It becomes clear that an automatic code generation minimizing the worst-case execution time is highly desired. Its application would not only significantly reduce the time-to-market period but would also allow the utilization of cheaper hardware for the highly optimized code. In this book, the drastic lack of methods for an automatic minimization of the WCET estimation of embedded software is addressed. The proposed WCET-aware compiler automatically reduces the program's WCET, thus completely removes the current tedious and error-prone trial-and-error based design. The idea behind WCET-aware compilation as well as its challenges are presented in the following.

1.1.2 WCET-Aware Compilation

With the increasing complexity of embedded software, high code quality can only be achieved with an optimizing compiler. The vital role of compilers is continuously growing with the importance of embedded software. For example, it has been estimated that embedded software will account for approximately 40% of total development costs in a car by 2010 [HKK04].

State-of-the-art compilers offer a vast spectrum of optimizations with the objective to minimize the average-case execution time or energy dissipation like e.g., the *encc* compiler [SW+10]. To enable a WCET-aware code generation, the compiler requires a tight integration of a static timing analyzer which estimates the WCET of the program under analysis. The invocation of the WCET analyzer should be performed automatically by the compiler in order to relieve the user from the burden of a manual configuration and execution of the analyzer. Afterwards, an automatic import of WCET data should take place to establish a WCET timing model within the compiler that can be exploited for optimizations.

The WCET awareness in the compiler offers a significant advantage. Any transformation to the code that is executed on an advanced target architecture is highly non-intuitive regarding its impact on the timing. This leads to the well-know problem that optimizations performed by current compilers may end up in a degraded system performance [ZCS03, CFA+07, LOW09]. On the contrary, the existence of a timing model enables an evaluation of code modifications and an avoidance of adverse effects.

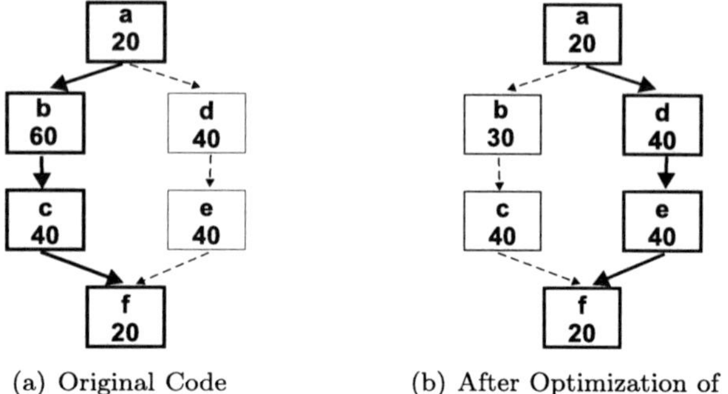

Fig. 1.3 Worst-case execution path switch

<table>
<tr><td align="center">(a) Original Code</td><td align="center">(b) After Optimization of **b**</td></tr>
</table>

1.1.2.1 Challenges for WCET Minimization

Compared to traditional compiler optimizations, WCET-aware optimization imposes novel challenges to the compiler. The well-known rule "make the common case fast" [HP03] is not applicable to real-time systems since infrequently executed portions of the code may significantly contribute to the program's WCET, turning them into promising optimization candidates.

To determine the crucial parts of the code, the compiler has to find the *worst-case execution path* (*WCEP*) which is the longest path through the program. Its length corresponds to the WCET and its shortening is equivalent to the reduction of the program's WCET. Other parts of the code that are not lying on the WCEP do not contribute to the worst-case performance and are not suitable for optimization.

For an effective WCET reduction, it is, however, not sufficient to compute the WCEP once and conduct all optimization steps on this initial data. The reason is the instability of the WCEP, also known as the *WCEP switch*. Figure 1.3 illustrates this issue with a control flow graph (CFG) where nodes represent basic blocks with their corresponding WCETs and edges denote control flow dependencies. Bold edges mark the WCEP. Starting with the CFG in Fig. 1.3(a), a transformation may optimize block b by 30 cycles. As a result, the WCEP switches to the right branch of the CFG as seen in Fig. 1.3(b). However, the overall WCET of 140 cycles did not decrease by 30 cycles but only by 20 as the length of the new WCEP amounts to 120 cycles. Optimizations have to be aware of the switch to avoid further transformations on the outdated WCEP.

To sum up, the challenges that a WCET-aware compiler has to face are the following:

- **Existence of a Precise WCET Timing Model**

    At any point during the optimization process, the compiler requires a detailed knowledge of the temporal behavior of the program under analysis. The exploited WCET timing models steer a systematic WCET minimization.
- **Novel Optimization Paradigms**

    Traditional compiler optimizations that predominantly aim at an optimization of the average-case performance may be not suitable for WCET minimization. Hence, either their adaption or even a complete redesign becomes inevitable.

- **Awareness of WCEP Switches**
  Code modifications necessitate an update of WCET information to warrant that optimizations do not operate on an outdated WCEP.
- **Multi-objective Optimizations**
  To suit multiple requirements imposed on the design of modern embedded real-time systems, a sophisticated compiler should provide the infrastructure for multi-objective optimizations, comprising the crucial cost functions WCET, ACET, and code size.

The previous discussion has emphasized the importance of optimizing compilers to provide the industry with embedded systems that satisfy both high efficiency requirements and stringent timing constraints. The following section provides an overview of the contributions of this book towards the optimization of embedded real-time systems.

1.2 Contribution of This Work

Conventional methodologies are not capable of assisting system designers during WCET-aware code generation. To overcome the current lack of tools, this work proposes a compiler framework, analyses, and a wide range of optimizations for WCET minimization. In detail, the following contributions are made:

- **WCET-aware Compiler Framework**
  The developed WCET-aware C compiler framework *WCC* represents a reconciliation of a sophisticated compiler and a timing analyzer. WCC is the first and currently only fully functional WCET-aware compiler (see overview of related work in Sect. 3.2). It is compliant with the ANSI C standard. Its target architecture is the Infineon TriCore TC1796 processor [Inf08a], a state-of-the-art architecture widely employed in the automotive industry. Due to its architectural features, such as caches and superscalar pipelines, code transformations can be evaluated in a realistic environment. This is in contrast to other ad-hoc compiler frameworks used by the research community—these frameworks use simple processors without caches and a tight integration of a WCET analyzer is not available. Extensions to the compiler framework that were developed in the course of this book will be explicitly pointed out in the respective sections.
- **Static Loop Analyzer**
  The infrastructure of the WCC framework has been extended by a static loop analysis which was developed in the course of this book. The loop analysis computes loop iteration counts automatically, representing mandatory information for a static WCET analysis. This way, the user is relieved from an error-prone and tedious specification of program characteristics turning WCC into a fully automated development tool.
- **Study of Impact of Standard Optimizations on WCET**
  Due to the current lack of suitable tools, the potential of code transformations on the temporal behavior of programs is only sparsely dealt with in today's literature. Using WCC's tight binding to a WCET analyzer, the impact of traditional

ACET optimizations on the program's worst-case behavior is explored. The book indicates that effects of standard optimizations may considerably differ for both objectives, the ACET and WCET.

- **Novel WCET-aware Source Code and Assembly Level Optimizations**

  The integration of a detailed WCET timing model enables the development of novel WCET-aware source code and assembly level optimizations. The book presents numerous optimizations that are derived from standard optimizations but have either been extended by WCET concepts or completely redesigned to better suit a WCET minimization. To exploit the high optimization potential of compiler optimizations at the source code level, detailed WCET data has to be imported from the compiler backend into the compiler frontend. For this purpose, the book proposes a technique called **back-annotation**. In addition, the book presents a paradigm called **invariant path** which helps to accelerate compiler optimizations suffering from worst-case execution path switches.

- **Machine Learning Based Compiler Heuristics for WCET Reduction**

  The development of heuristics for compiler optimizations which efficiently reduce a given objective for a broad range of applications is a tedious task that requires both a high amount of expertise and an extensive trial-and-error tuning. Machine learning has shown its capabilities for an automatic generation of heuristics used by optimizing compilers. The advantages of these heuristics are that they can be easily adopted to a new environment and often outperform hand-crafted compiler optimizations. In this book, supervised learning approaches are studied for the first time in the context of an automatic minimization of the WCET. A machine learning tool is integrated into the WCC framework and exploited for an automatic generation of machine learning based compiler heuristics for WCET reduction.

- **Multi-objective Exploration of Compiler Optimizations**

  The quality of the code significantly depends on the choice of compiler optimizations and their order. Modern compilers provide a vast portfolio of optimizations which exhibit complex mutual interactions and affect different design objectives in a hardly predictable fashion. This book is the first to explicitly address the multi-objective nature of embedded system design at compiler level by presenting a multi-objective exploration of compiler optimizations. Using evolutionary algorithms, the WCC framework is exploited for the search of optimization sequences with conflicting goals. The found solutions represent Pareto front approximations and indicate which trade-offs between the objectives WCET, ACET, and code size can be achieved. Providing these automatically generated solutions, the system designer is offered an overview of feasible solutions which help to find the optimization sequence that best suits the system requirements.

All of the proposed optimizations are implemented in the WCC compiler. The effectiveness of each optimization is demonstrated on a large number of real-life benchmarks, showing significant improvements for the considered optimization goals. As the presented optimizations are neither restricted to a particular programming language nor one specific target architecture, they can be easily adapted to other environments. Therefore, the overall contribution of this book is to offer researchers, embedded system designers and compiler writers a practical guideline for the optimization of real-time systems.

1.3 Outline

The remainder of this book is organized as follows:

- Chapter 2 gives an overview of existing timing analysis techniques, introduces basic concepts encountered with the static WCET analysis, and sketches the workflow of aiT, the timing analyzer employed in this book.
- Chapter 3 discusses the infrastructure of the WCET-aware compiler with a special emphasis on the integration of the timing analyzer, the static loop analysis, and the back-annotation.
- Chapter 4 presents different WCET-aware source code level optimizations as well as the invariant path paradigm.
- Chapter 5 discusses WCET-aware assembly level optimizations which are processor-specific, i.e., they exploit particular architectural features.
- Chapter 6 is dedicated to the automatic generation of machine learning based compiler heuristics for WCET minimization at both the source code and assembly level.
- Chapter 7 presents the multi-objective exploration of WCC's standard source code and assembly level optimizations for embedded real-time systems.
- Chapter 8 concludes the book and gives directions to important areas for future work.

Chapter 2
WCET Analysis Techniques

Contents

2.1 Introduction

Besides functional correctness, hard real-time systems have to satisfy timing constraints. They can be validated by a schedulability analysis that tests whether all tasks of the real-time system executed on the given hardware satisfy their timing constraints. The schedulability analysis requires safe and precise bounds on the execution time of each task. In this chapter, common timing analysis techniques found in academia and industry are discussed. Section 2.2 introduces the three main classes of approaches for a WCET analysis. The static timing analysis is the only method that guarantees safe WCET estimates. It does not rely on the execution of the code under analysis but derives its timing information from the program code combined with abstract models of the hardware architecture. Section 2.3 defines basic terms involved in the static analysis. Finally, the workflow of the leading static WCET analyzer *aiT* [Abs10], which is also exploited in this work, is presented in Sect. 2.4.

P. Lokuciejewski, P. Marwedel, *Worst-Case Execution Time Aware*   13
Compilation Techniques for Real-Time Systems, Embedded Systems,
DOI 10.1007/978-90-481-9929-7_2, © Springer Science+Business Media B.V. 2011

2.2 Approaches for WCET Analysis

The general classification of the approaches for a WCET estimation is based on the distinction whether the considered task is executed or statically analyzed.

2.2.1 Measurement-Based Approach

This method executes the task on the given hardware or in a simulator and measures the execution time. Representative sets of input data are provided which are supposed to cover scenarios where the maximal program execution time can be measured. This method has two main disadvantages.

First, it is not safe since the input to the program leading to the worst-case behavior is in general not known. To guarantee that the WCET is measured, the program must be executed with *all* possible input values which is not feasible in practice. Second, the measurement often requires an instrumentation of the code, e.g., augmentation with instructions to control hardware timers. However, the certification of safety-critical systems often specifies that the validation is carried out on exactly the same code that is utilized in the final product. For example, in the avionic domain it is mandatory to use the same code for validation that is also later used in the airplane (*Test what you fly and fly what you test* [The04]). Hence, the measurement-based approach is not applicable in such environments.

Measurement-based approaches are currently the most common techniques found in industry since the hardware under analysis or the respective simulators are usually available. However, as there is no guarantee that the maximal program run time was measured, a safety margin is added to the measured execution times which often leads to highly overestimated timing results [SEG+06].

2.2.2 Static Approach

The static approach emphasizes the *safety* aspect and produces bounds on the execution time which are guaranteed to be never exceeded by the execution time of the program under analysis. Unlike the measurement-based approach, the static approach considers all possible input values to the task. To scale down the complexity of an exhaustive analysis of all values, the large number of possible input data is reduced using a safe abstraction. In addition, the static approach does not execute the code on real hardware or a simulator but analyzes the set of possible control flow paths through the program. Using abstract models of the hardware architecture, the path with the maximal execution time can be determined. This longest path is called the worst-case execution path (WCEP) and its length corresponds to the program's WCET. The success of the static approach highly depends on the abstract hardware models. If they are correct and specify the underlying system precisely, safe upper bounds on the execution time of a program can be defined. However, it is in general hard to verify the correctness of the abstract models. Another drawback of the static analysis is that it might produce overestimated results if conservative decisions due to a lack of information during the analysis have to be taken. Furthermore, complexity of the static approach might be an issue if large programs are analyzed, leading to high analysis run times.

Currently, the static approach can be mainly found in academia. However, industry is recognizing the needs for safe WCET estimations. The growing industrial interest led to first applications of static WCET analysis in the automotive [BEGL05, SEG+06] and avionics [HLS00, FHL+01, SLPH+05] domain.

2.2.3 *Hybrid Approach*

The idea behind hybrid approaches is to combine concepts from the measurement-based and static approach. The hybrid approach identifies so-called *single feasible paths* (*SFP*) which are program paths consisting of a sequence of basic blocks where the execution is invariant to input data. To find SFPs at source code level, symbolic analysis on abstract syntax trees can be used [Wol02]. In the next step, the execution time of the SFPs is measured on real hardware or by cycle-accurate simulators. For input-dependent branches, input data for a complete branch coverage must be supplied. The execution time of these parts is also determined by measurements. In order to cover potential underestimation during the measurement, an additional safety margin is added to the measured execution time. Finally, the information of the SFPs is combined with techniques from the static approach to determine the longest path. The advantage of the hybrid approach is that it does not rely on complex abstract models of the hardware architecture. However, the uncertainty of covering the worst-case behavior by the measurement remains since a safe initial state and worst-case input can not be assumed in all cases. Moreover, instrumented code is required which may not be allowed in particular certification scenarios. This approach is used in the analysis tool suite *SymTA/S* [Sym10].

2.3 Basic Concepts for Static WCET Analysis

As shown in the previous section, the static approach is the only method to compute safe upper bounds on the execution time of a program. For compiler-based optimizations aiming at an automatic WCET minimization, reliable worst-case timing information is compulsory to achieve a systematic WCET reduction. Otherwise, unreliable timing information may result in adverse optimization decisions. This is the main reason why the WCET data exploited by the developed optimizations is computed statically.

Using the static approach for compiler optimizations is further motivated by practical reasons. First, many WCET-aware compiler optimizations operate iteratively, requiring multiple updates of the worst-case timing information. The measurement-based approach does not provide the desired flexibility for a frequent WCET estimation. Second, both the static WCET estimation and compiler optimizations operate on similar code and data representations enabling a tight exchange of information between both approaches.

In the following, basic terms found in the context of static WCET analysis are introduced. Moreover, specific problems and requirements for this class of timing analysis are discussed.

2.3.1 Control Flow

A static WCET analysis requires information on the hardware timing such as the execution time of individual instructions. Thus, the analysis must be performed on the assembly or machine level of the code. In addition, to statically derive a timing bound of the program, its possible execution flows must be known. This information is provided by a control flow analysis which operates on assembly *basic blocks* [Muc97]:

Definition 2.1 (Basic block) A *basic block* is a maximal sequence of instructions that can be entered only at the first instruction and exited only from the last instruction.

As a consequence, the first instruction of a basic block may be the entry point of a function, a jump target, or an instruction that is immediately executed after a jump or return instruction. Call instructions represent a special case. Typically, they are not considered as a branch since the control flow continues at the immediate successor of the call instruction after the callee was processed. Thus, basic blocks may span across calls. However, if it can not be guaranteed that the control flow continues immediately after the call instruction, the call must be considered a basic block boundary. Examples are calls in the Fortran language with alternate returns or ANSI C exception-handling mechanisms, like `setjmp()`/`longjmp()`, which force the control flow to continue at a point in the program other than the call's immediate successor. Since the original source code language might be unknown at assembly level, timing analyzers must be conservative, thus handling call instructions as basic block boundaries.

Basic blocks can also be defined at other abstraction levels of the code. For example, high-level basic blocks defined at source code level represent sequential code consisting of statements.

A common data structure to represent the program's control flow is given by the *control flow graph (CFG)*:

Definition 2.2 (Control flow graph) A *control flow graph* is a directed graph $G = (V, E, i)$, where nodes V correspond to basic blocks and edges $E \subseteq V \times V$ connect two nodes $v_i, v_j \in V$ iff v_j is executed immediately after v_i. $i \in V$ represents the start node, called *source*, which has no incoming edges: $\nexists v \in V : (v, i) \in E$.

In literature, a CFG typically represents the control flow within a single function. On the contrary, an *interprocedural control flow graph (ICFG)* [LR91] models the control flow of an entire program across function boundaries. It constitutes the union of control flow graphs G_f for each function f augmented by call, return, entry, and exit nodes. Call nodes are connected to the entry nodes of functions they invoke, while exit nodes are connected to return nodes corresponding to these calls.

A possible execution path π through the ICFG is defined as follows:

Definition 2.3 (Path) A *path* π through the control flow graph $G = (V, E, i)$ is a sequence of basic blocks $(v_1, \ldots, v_n) \in V^*$, with $v_1 = i$ and $\forall j \in 1, \ldots, n - 1 :$ $(v_j, v_{j+1}) \in E$.

Using the definition of a path, the term *program* is defined as follows:

Definition 2.4 (Program) All possible paths through the control flow graph $G = (V, E, i)$ starting at i and ending in a sink constitute a *program* $\mathcal{P}$. A sink represents a block $s \in V$ with $\nexists v \in V : (s, v) \in E$.

The program path with the maximal length, which is expressed by the number of execution cycles when this path is executed, is referred to as the worst-case execution path (WCEP).

2.3.2 Processor Behavioral Analysis

The task of the *processor behavioral analysis* is the determination of timing semantics of the hardware components that influence the execution time of the program under analysis. In particular, processor components such as different memories (caches, scratchpads, etc.), pipelines, and branch predictions must be taken into account. The result of this analysis is an upper bound for the execution time of each instruction or basic block.

The processor behavioral analysis relies on an abstract model of the target architecture. For a sound approximation, it is not mandatory to model all of the processor's functionalities precisely. Simplified models may be sufficient if they can guarantee that timing semantics are handled in a conservative way, i.e., the worst-case timing is never underestimated.

The static WCET analysis of simple processors without caches and pipelines allows a separate consideration of instructions. To estimate the WCET of the program, the WCET of each basic block is computed by accumulating the execution times of its instructions [Sha89]. The consideration of instructions independent of their *execution history* is no longer valid for modern processors and may result in a WCET estimation which is not safe. The reasons are context-dependent execution times of instructions and timing anomalies.

2.3.2.1 Context-Dependent Timing Analysis

For modern processors, the execution of instructions depends on its *context*, i.e., which instructions were previously executed. Such situations may arise when a basic block *b* with more than one predecessor is executed. Depending on the previously executed block, the cache or pipeline behavior during the execution of *b* may vary, leading to different WCETs. Another example are loops. The first loop execution may have a higher execution time than the following executions since data must be initially loaded into the cache, leading to *compulsory cache misses*. Thus, a sophisticated WCET analyzer must exploit the knowledge of the execution history to estimate context-dependent upper timing bounds for instructions.

2.3.2.2 Timing Anomalies

Due to the complexity of hardware and software, timing analysis often only becomes feasible when abstraction is introduced. Abstraction allows the handling of unknown input data and enormous spaces of processor states but comes at the cost of less precision. Unknown parts of the execution states lead to non-determinism if decisions during the analysis rely on this unknown information [RWT+06]. The separate consideration of an instruction may lead to different execution times depending on different assumptions about the missing information. Intuitively, it would be assumed that a locally faster execution entails a decrease of the overall program's WCET. However, on modern processors this does not need to be true. A *timing anomaly* is defined as follows [Lun02, The04]:

Definition 2.5 (Timing anomaly) If the change ΔT_i of the overall execution time of an instruction i results in a change $\Delta T_\mathcal{P}$ of the global execution time of program $\mathcal{P}$, a *timing anomaly* occurs if either

- $\Delta T_i < 0 \rightarrow \Delta T_\mathcal{P} < \Delta T_i \vee \Delta T_\mathcal{P} > 0$, i.e., the acceleration of i leads to a smaller acceleration of the program's execution time or the program runs longer than before,
- $\Delta T_i > 0 \rightarrow \Delta T_\mathcal{P} > \Delta T_i \vee \Delta T_\mathcal{P} < 0$, i.e., the program's execution time is more extended than the delay of instruction i or the program execution is accelerated.

The crucial case for a WCET estimation is $\Delta T_i < 0 \rightarrow \Delta T_\mathcal{P} > 0$. It points out that a consideration of a local worst-case scenario at instruction i is not sufficient for a safe estimation of the WCET. Since verifying the absence of timing anomalies is provably hard, timing analyzers are forced to consider all possible scenarios, i.e., to follow execution through several successor states whenever a state with a non-deterministic choice between successor states is detected. This may lead to a *state explosion*. A first approach towards a more efficient WCET analysis, which exploits precomputed information about the abstract model of the hardware in order to safely discard analysis states, was presented in [RS09].

For modern processors, timing anomalies due to three different processor features were observed. *Speculation timing anomalies* arise when costs of an expensive branch mis-prediction exceed the costs of a cache miss [HLT+03]. *Scheduling timing anomalies* occur when dependent instructions are differently scheduled on the hardware resources, e.g., pipelines, leading to varying execution times of instruction sequences [Gra69, Lun02]. Therefore, scheduling anomalies can be observed on *out-of-order architectures*. But even *in-order architectures* may show timing anomalies. For example, on the Motorola ColdFire 5307 processor with its pseudo-round robin cache replacement strategy, an empty cache may not constitute the worst possible cache behavior w.r.t. the execution time [The04].

2.3.3 Flow Facts

The static WCET estimation is based on static program analyses which attempt to determine dynamic properties of the program under analysis without actually executing it [CC77]. Some of these properties are not decidable in general. In particular, the determination of the longest path requires the knowledge of loop iteration counts. However, a static loop analysis is generally not decidable for a concrete program semantics since it includes the proof of termination. Thus, an automatic loop analysis is only feasible for a restricted set of programs [EE00]. To enable an automatic timing analysis of arbitrary programs, the user is typically forced to support the WCET analyzer with program annotations about the loop iteration counts. In a similar way, recursion depths and target addresses of dynamic calls and jumps, which can not be statically determined, make a user annotation mandatory. If the program execution is dependent on input data, e.g., in sorting algorithms, the user annotations must respect all potential values that may be provided to the program. In literature, information about the dynamic behavior of a program is referred to as *flow facts* [Kir03].

Definition 2.6 (Flow facts) *Flow facts* give hints about possible paths through the control flow graph of a program $\mathcal{P}$. Flow facts can be expressed implicitly by the program structure or semantics as well as by additional information provided by user annotations.

2.3.4 Bound Calculation

In literature, three popular approaches to compute upper bounds on the execution time of a program $\mathcal{P}$ based on its (interprocedural) control flow graph are presented. All these approaches require additional control flow information about loop iteration counts and recursion depths which can be either provided by an automatic analysis or manually annotated flow facts.

The *path-based* approach [SEE01] models each possible path in the ICFG explicitly. For each of these paths, the path length is computed. The length of the longest path reflects the WCET. The main drawback of this approach is its complexity since the number of possible paths grows exponentially with the number of control flow branches. Therefore, heuristic search methods are possibly required. The *tree-based* approach [Sha89, CB02] traverses the abstract syntax tree of a program at source code level in a bottom-up manner and computes upper timing bounds for connected code constructs. The computation of the timing bounds is steered by predefined combination rules which state how the execution times of parent constructs are derived based on the execution times of their child constructs. After the computation, the current construct is collapsed and its WCET is propagated to its parent constructs. This approach makes an integration of flow facts difficult and relies on a clear mapping between the source and machine code which is not always given for optimized code.

Another approach, which is nowadays widely used, is the *implicit path enumeration technique (IPET)* [LMW95]. In this approach, the ICFG of a program is translated into a system of linear constraints. According to Kirchhoff's rules, the following constraints are generated:

$$\forall v, v' \in V, \quad \{(v, v') \in E\} : c(v) = \sum_{(v,v') \in E} trav(v, v') \wedge$$

$$\forall v, v' \in V, \quad \{(v', v) \in E\} : c(v) = \sum_{(v',v) \in E} trav(v', v)$$

$c(v)$ represents the execution count of block v and $trav(v, v')$ reflects the edge traversal count of edge $e = (v, v')$. Using these constraints, the control flow can be modeled as an *integer linear program (ILP)* [Chv83]. The equations warrant that the traversal counts of all incoming edges of v are equal to the traversal counts of all outgoing edges of v.

The problem to be solved is to maximize the overall flow through the ICFG. For this purpose, an objective function expressing the estimated WCET $WCET_{est}$ of the program $\mathcal{P}$ is formulated:

$$WCET_{est} = \sum_{v \in V} t(v) \cdot c(v)$$

$t(v)$ represents the WCET of basic block v computed by the processor behavioral analysis. This sum is a linear combination since the execution times of the blocks are constants in the integer linear program. Using the sum as objective function to the ILP as a maximization problem yields the estimated WCET of $\mathcal{P}$, while the results for $c(v)$ indicate how often v will be executed on the WCEP. For a more precise analysis, the model can be extended by execution contexts.

IPET is a flexible approach since it allows an easy addition of complementary constraints about the control flow, such as information about mutually exclusive paths that increase the precision of the upper bounds. Unlike the path-based approach, IPET does not explicitly model paths, thus the order of blocks lying on the

Fig. 2.1 Workflow of static WCET analyzer aiT

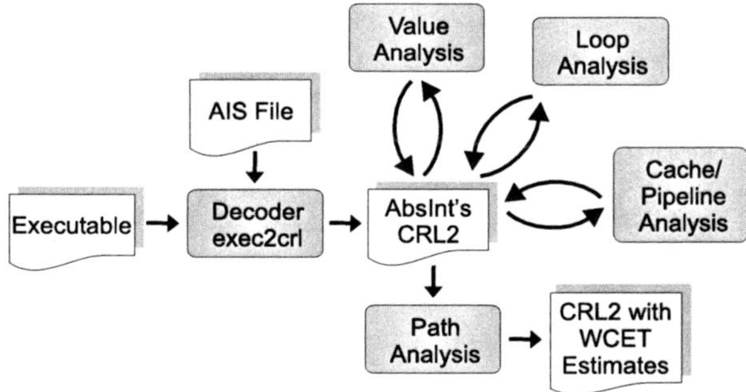

WCEP is not apparent. Despite the fact that solving of integer linear programs is $\mathcal{NP}$-complete, many large ILP problems can be solved in practice with a moderate amount of effort. For example, the WCET analysis provided by the tool aiT performed on an Intel Xeon 2.4 GHz machine takes 34 seconds for a simple cyclic redundancy check (CRC) computation [MWRG10] or 181 seconds for the encryption of a 32 bit message employing the commonly used secure hash algorithm (SHA) [GRE+01].

2.4 Static WCET Analyzer aiT

A leading static WCET analyzer is the tool aiT that is developed by the company AbsInt [Abs10]. Since it is used in this work for the calculation of upper timing bounds, its workflow will be briefly discussed in the following.

The modular architecture of aiT is depicted in Fig. 2.1 and consists of the following steps:

- As input, the timing analyzer expects a binary executable. It is disassembled by the tool *exec2crl* in order to construct an interprocedural control flow graph using the human-readable intermediate format *CRL2* (*Control Flow Representation Language*) [The02]. Moreover, flow facts from the parameter file called *AIS* might be imported to annotate the constructed ICFG with additional information. The initial CRL2 file is passed to the subsequent stages of aiT which annotate the file with further analysis results.
- The *value analysis* is based on an interval analysis to determine a sound approximation of potential values in the processor registers occurring at any possible program point. The values of the registers are exploited for the computation of possible addresses used for memory accesses. Moreover, the register values are exploited for the detection of *infeasible paths*, i.e., mutually exclusive paths that are never executed simultaneously in the same context.
- The static *loop analysis* tries to automatically determine the number of loop iterations for each loop in the ICFG. The analysis is based on pattern matching and involves results from the value analysis. aiT's loop analyzer succeeds for simple loops. Hence, user annotations for the remaining loops are required.

- The integrated *cache/pipeline analysis* relies on the register value ranges from the value analysis. The cache analysis tries to statically precompute cache contents. It performs a *must/may analysis* to classify memory references into sure cache hits and potential cache misses [FW99, HLT+03]. The pipeline analysis uses an abstract model of the target architecture to simulate the behavior of the pipeline [The04]. Considering timing anomalies and integrating results from the cache analysis, the pipeline analysis derives upper bounds on the execution time for each basic block depending on its context.
- The *path analysis* computes upper bounds for the overall program execution time. The computation is based on the IPET approach which exploits an ILP using the ICFG coded in CRL2 as well as timing results from the cache/pipeline analysis.

Chapter 3
WCC—WCET-Aware C Compiler

Contents

3.1 Introduction

Most embedded/cyber-physical systems have to respect timing constraints. System designers attempt to reduce the WCET of these systems since its reduction leads to a cut in the production costs: slower and cheaper processors can be used that

P. Lokuciejewski, P. Marwedel, *Worst-Case Execution Time Aware*   23
Compilation Techniques for Real-Time Systems, Embedded Systems,
DOI 10.1007/978-90-481-9929-7_3, © Springer Science+Business Media B.V. 2011

still satisfy the stringent timing constraints. Compilers play an essential role in the system design. Previous studies [Pua06, SMR+05, ZKW+05, FK09] have indicated the high potential for performance optimizations of embedded applications. Due to this reason, compilers promise to be an effective tool for an automatic WCET minimization.

Modern compilers fail to accomplish this goal since they are not aware of timing semantics. Current compiler optimizations mainly focus on the minimization of the average-case execution time. However, due to the lack of a precise timing model, the effects of optimizations can not be quantified, thus there is no guarantee for an improvement of the code performance. Even adverse effects on the transformed code can often be observed [CSS99, CFA+07, LOW09].

Another reason for the failure of current compilers is the almost unknown impact of ACET optimizations on the program's WCET. This uncertainty frequently leads to the common industrial practice of disabling all compiler optimizations during code generation of time-critical software.

In this chapter, the lack of code generation tools for embedded real-time systems is addressed. The WCET-aware C compiler WCC is proposed which enables an effective and automatic WCET minimization. The key feature of WCC is its tight coupling to a static WCET analyzer, the tool aiT (cf. Sect. 2.4), which allows an integration of a precise worst-case timing data into the compiler. Taking the notion of the program's worst-case timing behavior into account enables a detailed assessment of code transformations on the WCET on the one hand. On the other hand, the exploitation of the timing data provides a basis for the development of WCET-aware optimizations and analyses which aim at an aggressive WCET minimization.

The remainder of this chapter is organized as follows. Section 3.2 gives an overview of related work to existing frameworks that integrate WCET analyses. In Sect. 3.3, the basic workflow of WCC, which is similar to standard optimizing compilers, is presented to provide the reader with an overall overview of the WCET-aware compilation flow. The integration of a static WCET analyzer into the compiler framework is discussed in Sect. 3.4. In Sect. 3.5, modeling of flow facts is presented. A static loop analysis, which was developed in this book to automatically compute loop iteration counts, is discussed in Sect. 3.6. To exploit optimization potential for WCET minimization at source code level, WCET timing data has to be translated from the compiler's backend into its frontend. This technique, which was developed in this work, is called back-annotation and will be introduced in Sect. 3.7. Finally, an overview of WCC's target architecture, the Infineon TriCore processor, is provided in Sect. 3.8.

3.2 Related Work

A first approach to integrate a WCET analysis into a compiler was presented in [Bör96]. The framework expects flow facts in the form of source code pragmas which are forwarded to the compiler backend. Since no mechanisms to keep flow facts consistent during code transformation are supported, the application of compiler optimizations might invalidate the specified flow facts. Moreover, this framework uses a path-based WCET calculation which does not scale well for larger programs. Also, the target architecture is the Intel 8051 which is a simple and predictable processor since no cache, branch prediction, and pipeline are available.

In [Kir02], a prototype of a static timing analyzer for programs written in *wcetC* was presented. The language wcetC is a subset of C with extensions that allow to annotate the C code with control flow information that is exploited for the detection of infeasible paths. The tool cooperates with a C compiler that translates the source code into object code to perform the timing analysis. Due to missing automatic mechanisms for flow fact transformation, any structural code transformations by the compiler must be avoided. The supported platforms are the simple processors M68000, M68360, and C167.

The interactive compilation system called *VISTA* [ZKW+04] translates a C source code into a low-level IR used for code optimizations. The compiler framework integrates a proprietary static WCET analyzer supporting simple processors without caches like the StarCore SC100. VISTA is equipped with a loop analysis which is only able to detect simply structured loops. Code containing recursion can not be analyzed. The employed timing analyzer has a limited scalability, enabling the analysis of exclusively small program codes. Since the compiler lacks a high-level representation of the code, no optimization potential for WCET reduction at source code level can be exploited.

The open-source tool *Heptane* [CP01] is a static WCET analyzer with multi-target support for simple processors like StrongARM 1110 or Hitachi H8/300. It expects a C source code as input that is parsed into a high-level IR. Next, the code can be translated into a low-level IR. Heptane exclusively supports WCET-aware assembly level optimizations. The WCET can be computed either at source code level via a tree-based approach using combination rules for source code statements, or an ILP-based method that operates on a CFG extracted from the task's binary. Since the WCET analysis does not support a detection of infeasible paths, the derived upper bounds may be possibly considerably overestimated. Moreover, compiler optimizations are not supported during WCET analysis and must be disabled to avoid a mismatch between the syntax tree and the control flow graph.

SWEET [Erm03] is a static WCET analyzer with a research focus on flow analysis. It incorporates different techniques for the calculation of loop iteration counts and the detection of infeasible paths. Due to the missing import of WCET data into a compiler framework, the development of compiler optimizations aiming at a WCET minimization is not possible. The flow analysis is performed on a medium-level intermediate representation (MIR) of the code which is carried out after source code compiler optimizations. To avoid a mismatch between the MIR and the object code used for the WCET analysis, assembly level optimizations are not allowed. In addition, SWEET is coupled to a research compiler which is only able to process a subset of ANSI C [Sch00]. The supported pipeline analysis is limited to in-order pipelines and does not consider timing anomalies. SWEET's target architectures are the ARM9 and NEC V850E processor.

An integration of a static WCET analyzer into a compiler framework was presented in [PSK08]. The framework, called *TUBOUND*, allows the application of source code optimizations since flow facts specified as pragmas in the ANSI C code are automatically updated. This is achieved by extending supported optimizations by a mechanism that keeps flow facts consistent. This approach resembles the handling of flow facts in the WCC framework. However, in contrast to WCC's 21 flow fact aware source code optimizations, [PSK08] reports about three supported optimizations. Assembly optimizations are not available in TUBOUND due to a missing compiler backend support. Currently, the tool supports the simple C167 processor lacking caches and a pipeline.

Chronos [LLM+07] is an open-source static WCET analysis tool which translates C code into a binary executable and uses the binary code for the timing analysis. The main goal of Chronos is to estimate the WCET for modern processors with complex micro-architectural features. To this end, the WCET analysis supports the SimpleScalar simulator infrastructure and models its out-of-order pipeline, its dynamic branch prediction, and the processor's I-cache. The computed WCET data has been used in the Chronos framework for assembly level optimizations, such as a WCET-aware scratchpad allocation of data [SMR+05] or I-cache optimizations for body-area sensor network applications [LJC+10].

3.3 Structure of the WCC Compiler

In this section, the overall structure of WCC is presented. Some components of the compiler are discussed in more detail to provide the reader with a notion of how WCET-aware code generation and optimization is realized in WCC. Moreover, the following discussion indicates how the components that were developed in this book incorporate into the overall compilation flow.

The overall structure of WCC is depicted in Fig. 3.1. It consists of general components that can be found in ordinary optimizing compilers for the production of efficient assembly code. To support WCET-aware compilation, complementary functionality is required. The functionality comprises the generation and maintenance of flow facts, the production of input for the WCET analyzer, and the import of a worst-case timing data into the compiler. In the following, the workflow of WCC that resembles a traditional compiler design is introduced. These stages are connected by solid arrows in Fig. 3.1, while components turning WCC into a WCET-aware compiler are depicted with dashed arrows and will be discussed in the next sections.

3.3.1 Compiler Frontend ICD-C IR

The first step of the workflow is the parsing of ANSI C source files using the ICD-C compiler frontend [Inf10a], consisting of a parser, an *intermediate representation* (*IR*) of the code, and ACET source code optimizations. The parser is compliant

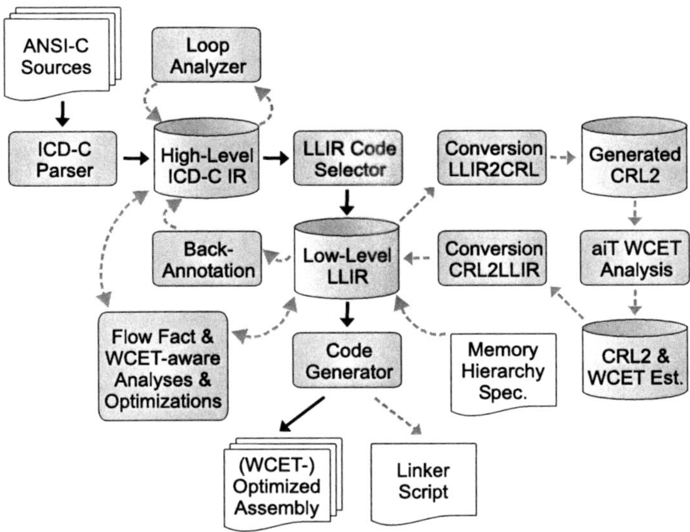

Fig. 3.1 Workflow of WCET-aware C compiler WCC

with the ANSI C standard. It comprises a lexical, a syntax, and a semantic analysis [ASU86]. The parsed code is finally translated into a high-level IR of the code. IRs are required for an automated code optimization and generation since a compiler requires internal data structures that model relevant characteristics of the program. Within the frontend, WCC uses a machine-independent high-level IR called *ICD-C IR*. Due to its early construction in the compiler chain, the IR is characterized by its closeness to the source language, enabling a back-transformation into the source code.

ANSI C constructs are represented by abstract data types. The simplified class model of the ICD-C IR is shown in Fig. 3.2. A program may consist of multiple source files represented by compilation units. The body of a function is modeled by a *top compound statement* which corresponds to code embedded in curly braces ({ }). The abstract statement class represents different statement types and may hold further statements or expressions which in turn may exhibit a hierarchical structure to allow the representation of nested expressions as shown in the following example.

Example 3.1 The **while**-loop statement holding an assignment expression with an array access on the right hand side is modeled by the ICD-C IR as follows:

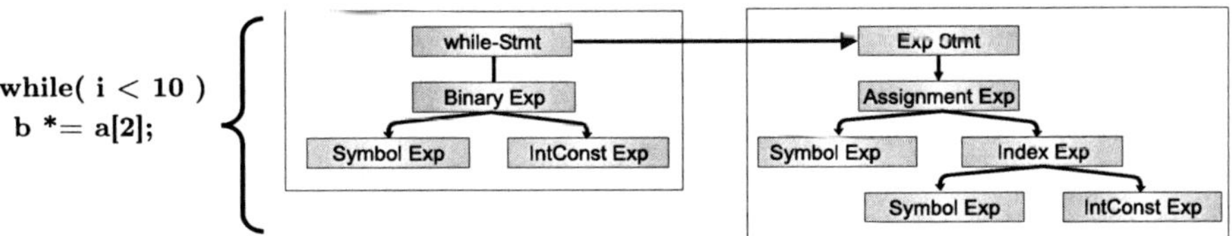

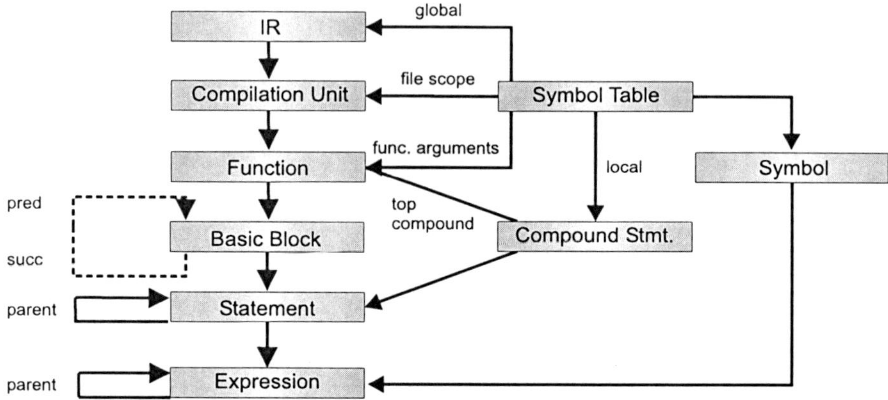

Fig. 3.2 Simplified class model of ICD-C IR

The assignment expression is a hierarchical expression consisting of a nested subtree for **a[2]**.

Moreover, as can be seen in Fig. 3.2, the statements within a function can be accessed via basic blocks. Within the WCC framework, exception-handling mechanisms, such as `setjmp()` (cf. Sect. 2.3.1), are not supported and must not be contained in the code under analysis. Due to this restriction it can be assumed that the control flow continues immediately after a call instruction. As a result, call statements in the ICD-C IR do not terminate a basic block.

To preserve the scope and data types of variables, different hierarchical symbol tables are involved. This allows to associate each identifiers with a particular scope, such as file scopes or local scope within a compound statement.

The ICD-C framework provides various control flow analyses based on the program's CFG, such as the determination of dominance relations or a reachability analysis. Moreover, data flow analyses, e.g., def/use chains, are available. For a detailed description of these standard techniques, the reader is referred to [Muc97].

In addition, ANSI C pragmas can be used to annotate the source code with user-defined information, e.g., flow facts about loop iteration counts that are required by the timing analyzer (like `_Pragma( "loopbound min 50 max 100" )`). In the parsing stage, this information is attached to the respective code constructs. This mechanism enables an efficient import of user data into the compiler frontend without the use of additional overhead such as external files. Another mechanism, which serves for storage of arbitrary information within the compiler, are so-called *persistent objects*. They represent generic containers that can be attached to arbitrary IR constructs. Within WCC, they are for example automatically filled with WCET data and assigned to the respective basic blocks. Assigning complementary cost functions enables the development of multi-objective optimizations.

Table 3.1 Available standard source code level optimizations

O1	Constant Folding
	Dead Code Elimination
	Local Common Subexpression Elimination
	Merge Identical String Constants
	Simplify Code
	Value Propagation
O2	Create Multiple Function Exit Points
	Life Range Splitting
	Loop Deindexing
	Loop Unswitching
	Optimize if-Statements in Loop Nests
	Remove Unused Function Arguments
	Remove Unused Returns
	Remove Unused Symbols
	Remove Unused Function Arguments
	Struct Scalarization
	Tail Recursion Elimination
	Transform Head-Controlled Loops
O3	Function Inlining
	Function Specialization
	Loop Unrolling

3.3.2 Standard Source Code Level Optimizations

To achieve a high code quality, compiler optimizations are essential. The ICD-C compiler framework is equipped with 21 standard ACET optimizations. Among others, they comprise data flow optimizations such as *dead code elimination* or *common subexpression elimination*, loop transformations such as *loop collapsing* or *loop unrolling*, and interprocedural optimizations like *function inlining* or *function specialization*. For details about the optimizations, the interested reader is referred to standard compiler literature [Muc97].

Compiler optimizations are typically classified into optimization levels. The higher the optimization level, the more optimizations are activated. An overview of the available source code optimizations in WCC subdivided into the optimization levels *O1*, *O2*, and *O3* is depicted in Table 3.1. Optimizations in *O3* are code expanding transformations which should be performed with caution if strict code size constraints are imposed on the system.

3.3.3 Code Selector

To lower the abstraction level of the code, the high-level ICD-C IR is translated into assembly code. For this purpose, a set of assembly operations has to be found which

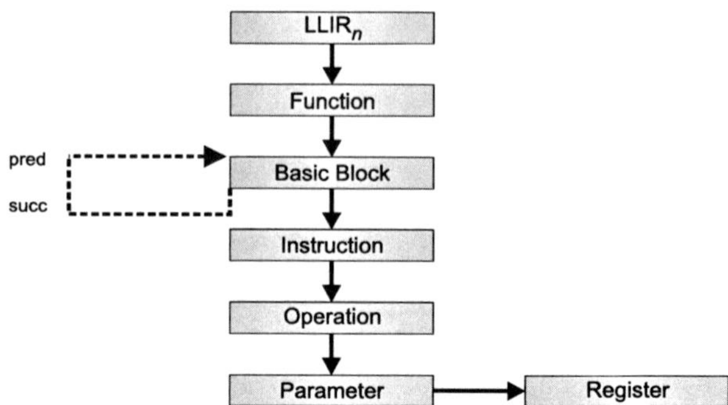

Fig. 3.3 Simplified class model of LLIR

represents a semantically correct translation of a given source code construct having minimal accumulated costs w.r.t. a given metric. The translation is carried out by a tree-pattern matching based *code selector* [FHP92] which accepts a cost-augmented tree grammar as input and emits an optimal parse of trees in the language described by the grammar. By parsing the trees in the found order, appropriate operations are generated—in case of WCC, TriCore assembly code is emitted.

3.3.4 Compiler Backend LLIR

The output of the code selector is an abstract representation of the processor-specific assembly code. The code representation used in WCC's compiler backend is the *low-level IR LLIR* [Inf10b]. It provides generic data structures for the abstraction of assembly code. Using a description of the target architecture, the retargetable LLIR turns into a processor-specific representation, allowing the exploitation of given processor features.

The hierarchical design of the LLIR enables an easy maintenance and manipulation of the program structure. A simplified class model of the LLIR is depicted in Fig. 3.3. An LLIR object is equivalent to a single assembly file and corresponds to a compilation unit within the ICD-C IR. If the compiled program consists of multiple source files, a list of LLIR objects has to be managed. Analogous to ICD-C, a program consists of functions and basic blocks. A basic block holds instructions which may consist of multiple operations. This capability allows to model bundled instructions such as required for e.g., VLIW architectures. For WCC's target architecture, the TriCore processor, an instruction holds exactly one operation. The operations are parameterized with constants, labels, architecture-specific operators, or registers. The latter are modeled by a separate class since they provide versatile information relevant for a compiler.

Similar to the ICD-C IR, the LLIR backend is equipped with various static analyses. A control flow analysis allows to iterate over the code structure according to the program's control flow. Moreover, relationships among basic blocks, such as reachability or dominance, are computed. In addition, data flow analyses comprising a livetime analysis and a def/use chain analysis [Muc97] are available.

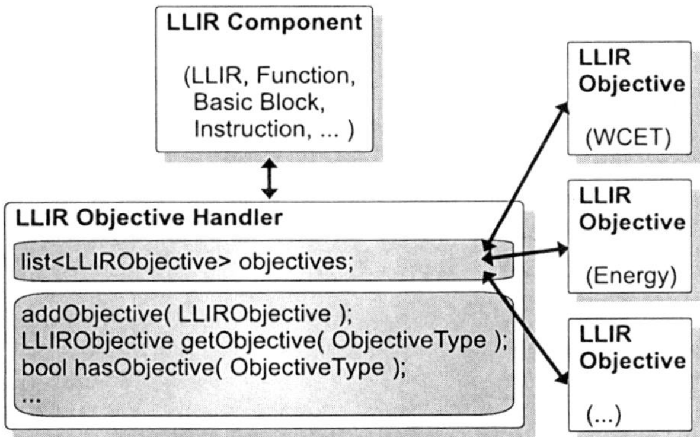

Fig. 3.4 LLIR objective handler

To annotate the assembly code with arbitrary data relevant for the compiler, LLIR provides the *objective* mechanism. Objectives are generic containers attachable to any LLIR component. The assignment of objectives to LLIR components is managed by a so-called *objective handler*. It is responsible for the adherence of consistency, e.g., disallowing the assignment of multiple objectives of the same type to one LLIR component, and provides methods to set and get objectives of a certain objective type. The general structure of this mechanism is depicted in Fig. 3.4. For a detailed description of this mechanisms, the reader is referred to [Lok05].

Besides the current use for WCET data, the integration of further objectives, such as energy consumption or ACET, is intended to allow the development of compiler optimizations pursuing different optimization goals.

3.3.5 Standard Assembly Level Optimizations

The assembly optimizations within WCC are predominantly processor-specific, i.e., they operate on the generic LLIR structures but exploit information about the Tri-Core architecture, e.g., the register file or the instruction set. The optimizations can be divided into two main classes: those that operate either on a *virtual* LLIR or a *physical* LLIR.

Using the code selector, the high-level IR of the code is typically translated into a virtual low-level IR. *Virtual* means that no physical registers but place holders identifying dependencies among operations are used. Many compiler optimizations benefit from virtual registers. These registers are not restricted in their number, thus provide a higher flexibility for the optimizations. The WCC compiler provides numerous standard ACET optimizations performed on the virtual LLIR including data flow optimizations, such as *constant propagation, loop invariant code motion*, or *peephole optimizations* [Muc97].

Another important optimization, which is mandatory for the generation of high-performance code, is *register allocation*. Its goal is to find a suitable mapping of

Table 3.2 Available standard assembly level optimizations

Virtual	Redundant Code Elimination
	Dead Code Elimination
	Constant Propagation
	Constant Folding
	Loop Invariant Code Motion
	Peephole Optimizations
	Register Allocation
Physical	Generation of 16 bit Operations
	Local Instruction Scheduling

virtual registers either to processor-internal registers or to memory. Since the number of physical registers is limited, the challenge for a register allocation is to map as many virtual registers to physical registers in order to avoid slow memory accesses. If more physical registers are required than available, some of the variables held in registers must be saved to memory in order to make the registers available for other variables. In the further program execution, the variables are restored back to registers. The shift of register values to and from memory is performed by load and store instructions, known as *spill code*, and should be avoided as much as possible. WCC's register allocation is based on graph coloring [Bri92]. The output of the register allocator is a physical LLIR.

WCC currently performs two optimizations on the physical LLIR. First, 32 bit operations are possibly converted into equivalent 16 bit operations. For example, the 32 bit TriCore operation `add d8, d8, d9` can be exchanged by the 16 bit operation `add d8, d9`, leading to an overall reduced code size. Moreover, local instruction scheduling operating on basic blocks [CT04] is applied. The optimization goal is to find an order of instructions which improves instruction-level parallelism yielding a reduced program execution time. WCC's scheduler focuses on the maximal utilization of the superscalar pipeline which can execute up to three operations in parallel. An overview of the available LLIR optimizations classified by the IR type is presented in Table 3.2.

3.3.6 Code Generator

The final step of the compiler is the code generator that emits valid TriCore assembly code from internal LLIR data structures. This code can be passed to an assembler/linker to generate the final binary executable. WCC utilizes the tools *tricore-as* and *tricore-ld* from the *TriCore GCC* tool chain developed by the company HighTec [Hig10] for this purpose.

3.4 Integration of WCET Analyzer

Precise timing data is computed by a static WCET analyzer using its internal worst-case timing model. To make timing information available in the WCC framework, a re-implementation of timing analyses from scratch within the compiler is not advisable. Rather, the development of complex WCET analyzers should be performed by timing experts while compiler developers should focus on code generation and optimizations. Following this methodology, the WCC compiler and the static WCET analyzer aiT are two separate tools that are tightly coupled in the compiler backend for a seamless exchange of information.

The integration of aiT into the compiler framework consists of three main steps: the translation of the program under analysis into a representation suitable as input for aiT, the transparent invocation of the timing analyzer, and the import of the computed worst-case timing information back into the compiler. These steps will be discussed in more detail in the following.

3.4.1 Conversion from LLIR to CRL2

The WCET analysis requires information on the hardware timing which is available on assembly/machine code level but not on higher abstraction levels of the code. For this reason, the timing analysis takes place in WCC's compiler backend. As mentioned in Sect. 2.4, aiT expects as input the program under analysis in the intermediate format CRL2 which is the IR of the timing analyzer. Hence, the first step of the integration of aiT is a conversion of LLIR code into CRL2. The conversion step replaces the decoder *exec2crl* from the ordinary workflow of aiT employed as a stand-alone tool on a binary executable (cf. Fig. 2.1 on p. 21).

It should be noted that aiT's decoder can not be used since the generation of the executable would require the application of an external assembler and linker which may, however, possibly modify the code in an irreproducible way, making an exchange of information between CRL2 and LLIR infeasible.

The conversion problem is equivalent to the general problem of program compilation. However, in contrast to compilers that aim at translating programs between (almost) arbitrary programming languages, WCC's CRL2 conversion translates programs that are represented at the same abstraction level. Thus, there is no *semantic gap* between LLIR and CRL2 which eases the conversion.

Since both LLIR and CRL2 are low-level IRs, a mutual translation of their CFGs is feasible. Both representations consist of functions, basic blocks, and instructions holding a single operation. Since CRL2 is usually constructed from a single binary file, the entire program must be represented by a single CFG. To this end, WCC conversion places all LLIR instances that were derived from various source files (compilation units) consecutively into a single CRL2 file.

Due to the analogy of both IRs, the conversion basically traverses the CFG of the LLIR recursively and generates equivalent CFG components for the CRL2 IR.

To enable an information exchange between the two representations, the conversion stage has to keep track of which CRL2 components were derived from which LLIR components. To enable a mapping between LLIR and CRL2 blocks, the *block conversion* function is introduced.

Definition 3.1 (Block conversion) Given two low-level intermediate representations $\mathcal{P}$ and $\mathcal{Q}$, a set of basic blocks $\mathcal{B}_P \in \mathcal{P}$ and a set of basic blocks $\mathcal{B}_Q \in \mathcal{Q}$. The bijective function $\texttt{blockmap} : \mathcal{B}_P \to \mathcal{B}_Q$ is called *block conversion*. The mapping uses the block label as a unique key.

Due to the bijection, the basic block labels used as key allow a mutual mapping between blocks of both IRs. Identifying first basic blocks of a function, which have block labels equal to the names of their surrounding function, makes a mapping between LLIR and CRL2 functions possible.

However, there are three issues that complicate the mutual translation. The first two problems stem from the fact that CRL2 is usually generated from a binary executable, relying on assembler and linker information. The last issue arises from structural transformations performed by aiT on the CFG which introduce a deviation to the LLIR CFG.

3.4.1.1 Operation Identification

In the ordinary workflow of aiT as a stand-alone tool, the first step consists of a control flow reconstruction based on a binary executable. The reconstruction includes an extraction of machine operations from a byte stream. The found sequences of bytes are subsequently matched against the processor vendor's machine code specification. This automatic approach is flexible since an exchange of the architecture merely requires a replacement of the machine specification.

An efficient algorithm for this purpose used in aiT was presented in [The00]. It is provided with a set of bit patterns—one for each machine operation to be recognized—and it recursively computes a decision tree for decoding. In contrast to a simple approach, checking each pattern in the set until a match was found, the tree-based approach can reduce the algorithm complexity from $O(n)$ to $O(\log(n))$, with n representing the size of the set of bit patterns.

The advantage of machine code decoding at machine level is that it can be reduced to an efficient bit pattern recognition. If the matching succeeds, the *mnemonic* (abbreviation for operation type) and the operation operands are unambiguously known. Within WCC, the program to be analyzed is represented by LLIR which is the basis for the construction of CRL2 code. The CRL2 code generation entails two problems.

First, to enable an efficient generation of CRL2 operations, AbsInt's code generation framework including the machine code specification should be exploited. Within the machine code specification, the specification of each operation o is defined by a unique identification number $\mathcal{I}_o$. Knowing $\mathcal{I}_o$ allows the extraction of detailed information about o and a generation of an appropriate CRL2 operation when further details about the operation parameters are available. Analyzing the byte stream bs from a binary delivers $\mathcal{I}_o$, thus machine code decoding can be considered as the mapping function map_{bs}:

$$map_{bs} : bs \mapsto \mathcal{I}_o$$

Furthermore, bs reveals the required information about the parameters. In contrast, an LLIR operation is represented by its mnemonic and a set of operands. Since assembler and linker information is missing, the operation identifier $\mathcal{I}_o$ is not available. To retrieve this crucial attribute, the machine specification must be exploited in a reversed manner: $\mathcal{I}_o$ is determined by finding a matching between LLIR characteristics and the machine specification. Hence, the problem can be formulated as follows:

Problem 3.1 (LLIR operation identification) Let C be a set of available characteristics $c \in C$ describing LLIR operation o_{LLIR}, and *SPEC* the machine specification. The problem of determining the corresponding identifier $\mathcal{I}_o \in SPEC$ for o_{LLIR} is to find a sufficient set $C' \subset C$, such that a bijective mapping can be established: $m_o : C' \mapsto \mathcal{I}_o$

Using all available characteristics C of an operation o_{LLIR} is inefficient since several characteristics are either irrelevant or redundant. For example, knowing if the first register parameter is written or read is not a significant information.

Likewise, extracting an insufficient set of characteristics may result in a surjective or injective mapping to $\mathcal{I}_o$. For example, the complex instruction set architecture (ISA) of the TriCore processor contains four different operations with the mnemonic AND [Inf08b]:

AND Dc, Da, Db	AND Dc, Da, *const9*
AND Da, Db	AND D15, *const8*

All these AND operations differ by the number and types of parameters, namely data registers (denoted by Dx or implicit register D15), or constants. As can be seen, the mere use of the mnemonic does not yield an unambiguous $\mathcal{I}_o$ in the machine specification. The distinction between TriCore's transport operations, like the word-wide load operations *LD.W*, is even more complicated since eight versions exist. To find a specific load operation, further characteristics, such as the addressing mode or the operation bit width, must be taken into account.

For the TriCore ISA, the operation identification algorithm has to take the following operation characteristics into account:

- mnemonic
- operation bit width (16 or 32 bit)
- number of parameters
- type of data/address registers: general or implicit (D15/A15)

- bit width of constants
- bit width of offsets (for load/store operations)
- addressing mode (for load/store operations)

The second problem emerging during the generation of CRL2 code concerns the adherence of consistency between the generated CRL2 code and the final binary executable generated from WCC's LLIR code. The problem is the internal behavior of an assembler that may translate ambiguous assembly operations differently based on internal rules that are often not sufficiently documented. A deviation between the CRL2 code provided as input for aiT and the generated machine code must be strictly avoided, otherwise the safeness of the estimated WCET is not guaranteed anymore. For instance, there are two versions of the MOV operation in the TriCore ISA:

MOV Dc, Db MOV Da, Db

The parameters of these two operations are arbitrary data registers in both cases. However, the first operation is 32 bits wide, whereas the second operation is a 16 bit operation. Within the LLIR, the bit width information is available and is also exploited for the generation of the respective CRL2 operation. In contrast, the assembler can optionally handle this operations either as 32 bit or 16 bit wide. To eliminate such uncertainties, the operation size has to be made explicit to the assembler. Within WCC, this is accomplished by the generation of directives [EF94] that are interpreted by the assembler, e.g.,

```
.code32
MOV D9, D8
```

This directive ensures that the assembler handles this operation as 32 bit wide.

3.4.1.2 Exploitation of Memory Hierarchy Specification

Another challenge for the conversion of LLIR into CRL2 are physical addresses. When CRL2 is constructed from a binary, the complete physical memory layout of the program is available. This includes the information about the memory utilized for the storage of code and data, i.e., the physical addresses of basic blocks and global data objects. Moreover, the assembler and linker resolve symbolic labels used in the assembly code for references to branch targets and global memory addresses into physical addresses. Based on this code, an adequate control flow graph for the determination of the longest path can be constructed. In addition, the knowledge about addresses involved in memory accesses to global data objects enables a precise WCET estimation.

To make a model of the physical memory layout available within WCC's backend, a *memory hierarchy specification* is provided (cf. Fig. 3.1). It resembles the commonly used *architecture description languages (ADLs)*, but provides only a simplified timing model of the processor's memory hierarchy. Rather, key parameters relevant for the generation of aiT's CRL2 input file and for compiler optimizations exploiting memory hierarchies are specified. A detailed memory timing model is integrated into the cache/pipeline analysis of aiT and is transparent to the user.

WCC provides a text file interface to specify the available processor memory hierarchies. The specification describes different regions of the processor's physical memory as well as parameters concerning TriCore's instruction cache [Kle08, Lok05]. The following attributes can be defined:

- a memory region's base address and absolute length
- access attributes, e.g., read, write, executable
- memory access times (relevant for optimizations)
- list of assembly sections allocatable to specific memory regions
- I-cache parameters, e.g., capacity, associativity, line size

A fragment of the memory hierarchy specification describing the SRAM data memory of the TriCore TC1796 processor is shown below:

```
# Data SRAM (DMU)
[DMU-SRAM]
origin = 0xc0000000
length = 0x10000 # 64K
attributes = RWA # read/write/allocatable
sections = .data.sram
```

Based on this information, the physical memory layout of code represented by the LLIR can be derived. It is utilized during the generation of CRL2 code for the determination of physical block addresses. In addition, branch targets of jump operations, which are represented by symbolic block labels, have to be translated into physical addresses. In a similar fashion, symbolic labels involved in accesses to global variables via load/store operations have to be converted.

Using the memory hierarchy specification, code and data can be allocated to different memory regions. These decisions need not only to be respected by the WCET analysis but also in the linkage stage for the generation of the binary. To accomplish a consistent code representation, WCC's code generator (cf. Fig. 3.1) is capable to write a suitable linker script that is passed to the finally invoked linker.

3.4.1.3 Loop Transformation

As mentioned previously, aiT performs structural transformations leading to the effect that the CFG structure of LLIR and CRL2 differ. The transformations aim at an increased precision for the WCET estimation. Section 2.3.2.1 indicated that the use of contexts yields an improved timing analysis since the execution history of instructions is regarded. Due to the structure of CRL2, context information can only

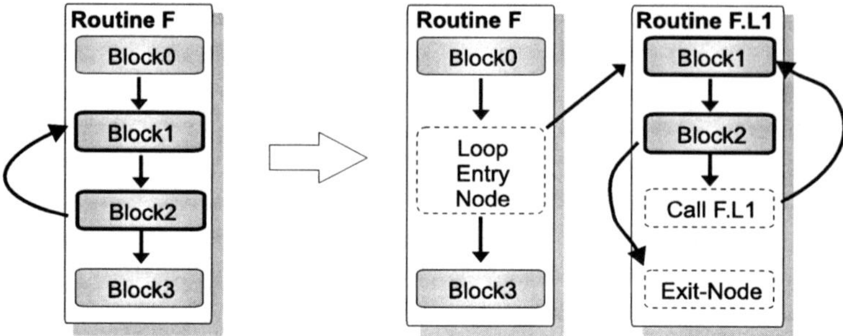

Fig. 3.5 Loop transformation

be attached to CRL2 functions. However, a precise timing analysis also has to distinguish between different loop iterations. For example, the cache content for the first iteration may differ from the content in the remaining iterations.

To enable a context-sensitive analysis of loops, aiT modifies the CFG using a loop transformation. The transformation introduces a new routine and a new call node for each loop and transforms the loop into a recursive routine [LT79, The02]. The loop transformation is illustrated in Fig. 3.5. Moving loop blocks *Block1* and *Block2* into the dedicated routine $F.L1$ facilitates the annotation of context information. Moreover, auxiliary nodes (marked by dashed lines) have to be inserted.

The mutual conversion between LLIR and CRL2 needs to be aware of the transformed CFG. When importing WCET data from the annotated CRL2 code back into LLIR, the newly created loop functions, such as $F.L1$, have to be distinguished from *real* CRL2 functions having a corresponding counterpart in the LLIR. The WCET data attached to these functions is stored within the LLIR at the corresponding first basic block representing a loop. For example, WCET data computed for loop $F.L1$ in Fig. 3.5 would be assigned to the LLIR basic block corresponding to *Block1*. The mapping between LLIR and CRL2 basic blocks is trivial by the means of the unique block labels in both IRs.

In addition, the application of the loop transformation affects the interpretation of WCET data related to CFG edges which is also imported into the LLIR. Since the LLIR does not explicitly model control flow edges, WCET data related to a CRL2 edge $e = (b_{source}, b_{target})$ is attached to the LLIR block corresponding to b_{source} with a reference to b_{target}. During the import, edges from/to the auxiliary CRL2 nodes have to be translated using the block conversion function from Definition 3.1. For example, WCC has to translate the source of the CRL2 edge starting at the loop entry node from this node to *Block2*.

3.4.2 Invocation of aiT

After the conversion of the code from LLIR into CRL2, the generated CRL2 file is automatically passed to aiT. The invocation of the timing analyzer is fully transparent to the user since the WCC framework steers the application of the required steps for the WCET analysis.

The compiler user benefits from the encapsulation of the timing analyzer into WCC. Instead of running a compiler to generate the binary, setting up the configuration parameters for the WCET analyzer, and finally manually invoking the analysis, all these steps are carried out by WCC in the background. Thus, the user is relieved from the burden of a manual configuration of a suitable run-time environment for aiT.

Another benefit of the encapsulation is that the compiler has a deep knowledge of the code that can be automatically exploited for an automatic computation of information which improves the precision of the WCET analysis. For instance, the timing analyzer can be provided with data about infeasible paths or statically known register values. This information is usually hard to compute manually and, if known anyway, its propagation to aiT would require a tedious and error-prone specification via a file interface.

3.4.3 Import of Worst-Case Execution Data

The last step of the integration of a timing analyzer into the compiler is the import of the worst-case timing data computed by aiT into the backend. For that purpose, the final CRL2 file, which represents the program's CFG enriched with WCET data, is traversed. Using the conversion function from Definition 3.1, a mapping between basic blocks of both IRs is established enabling an exchange of WCET data. The imported data is attached to the corresponding LLIR blocks in the form of LLIR objectives. The exchange of WCET data concerning functions works analogously. The following information (considered over all execution contexts) is made available within the LLIR after the import:

- worst-case execution times for the entire program, each function, and each basic block
- worst-case execution frequency per basic block and CFG edge
- worst-case call frequency per function
- execution feasibility of CFG edges
- safe approximation of register values
- computed I-cache misses per basic block

The approaches presented in this section were published in [FLT06a, FLT06b].

3.5 Modeling of Flow Facts

To make a static WCET analysis feasible, sufficient information about the execution counts of every instruction must be available. Some flow facts (cf. Sect. 2.3.3) can be automatically extracted from the structure of the program by means of static program analyses. However, due to undecidability it is in general impossible to calculate all required information about the control flow of a program. To overcome this problem, missing flow facts have to be provided as user annotations.

3.5.1 Specification of Flow Facts

The most convenient approach to specify user flow facts is the annotation of the program at source code level. Defining them at this high abstraction level rather than at the low level is more intuitive and less error-prone. Moreover, their direct annotation in the source code is recommended since only a single code base has to be maintained. The alternative approach of providing user-defined flow facts in separate files is inconvenient since they introduce additional overhead.

Within WCC, flow facts are inserted into the source code as ANSI C pragmas that are translated into internal data structures during parsing. The first class of flow facts are *loop bounds* limiting the iteration counts of regular loops, i.e., `for`-, `while`-, and `do-while`-loops with a single entry point and a well-defined termination condition.

Loop bound flow facts denote the minimum and maximum number of loop iteration counts for each loop and use the following EBNF syntax [Sch07]:

> **LOOPBOUND** |= loopbound min <u>NUM</u> max <u>NUM</u>
> **NUM** |= *Non-negative Integer*

Example 3.2 For the following code snippet with variable loop bounds that depend on the function parameter **arg** and the assumption that **func** was invoked with the arguments **10** and **100**, the user has to specify the loop bound pragma as:

```
void func( int arg ) {
 _Pragma( "loopbound min 10 max 100" )
 for ( i = 1; i <= arg; i++ )
   Statement block A
}
```

To annotate irregular loops and recursive function calls, *flow restrictions* are introduced. Within WCC, so-called *markers* are used to restrict the execution counts of statements relative to other statements:

> **MARKER** |= marker <u>NAME</u>
> **NAME** |= *Identifier*

Using markers that serve as a reference point within the program, the following EBNF grammar enables the formulation of relations between statements [Sch07]:

> **FLOWRESTRICTION** |= flowrestriction <u>SIDE</u> <u>COMPARATOR</u> <u>SIDE</u>
> **COMPARATOR** |= >= | <= | =
> **SIDE** |= <u>SIDE</u> + <u>SIDE</u> | <u>NUM</u> * <u>REFERENCE</u>
> **REFERENCE** |= <u>NAME</u> | *Function Name*

Example 3.3 To bound a triangular loop, a relationship between a statement **A** outside the loop and a statement **B** within the loop nest can be expressed by the following pragmas:

```
_Pragma( "marker outer" )
A;

for ( i = 0; i < 10; i++ )
  for ( j = i; j < 10; j++ )
    _Pragma( "marker inner" )
    B;

_Pragma( "flowrestriction 1*inner <= 55*outer" );
```

From the inequation it can be inferred that statement **B** is executed at most 55 times as often as statement **A**.

3.5.2 Translation and Transformation of Flow Facts

The final goal is to pass the flow facts specified at source code level to aiT. As a consequence, flow facts have to be successively lowered between the available IRs. The traversed chain of IRs including the deployed tools for the translation is shown below:

$$\text{ANSI C} \xrightarrow{\text{parser}} \text{ICD-C} \xrightarrow{\text{code selector}} \text{LLIR} \xrightarrow{\text{converter}} \text{CRL2}$$

In the last step representing the conversion from LLIR into CRL2, flow facts are also translated and modeled in CRL2. The subsequent WCET analysis exploits this data to formulate constraints for the ILP model of the IPET approach (cf. Sect. 2.3.4).

Besides the translation of flow facts, care must be taken when code is modified at any abstraction level. In particular, compiler optimizations might invalidate flow facts. For example, loop unrolling changes the number of loop iteration counts. These modifications must be respected to generate valid flow facts for aiT that reflect the effective CFG. The consistency of flow facts within WCC is preserved during code optimizations by extending the respective optimizations by individual *flow-fact update* techniques. They ensure that the maintenance of flow facts is simultaneously performed with code modifications.

3.6 Static Loop Analysis

The WCC compiler framework represents a tool for an automatic WCET minimization. The essential component is the static timing analysis which relies on flow facts. The manual annotation of flow facts in the source code as discussed in Sect. 3.5 is applicable for smaller programs for which the source code is available. The annotation of larger applications becomes tedious end error-prone. Moreover, this approach is not suitable for applications automatically generated by graphical specification tools.

In addition to the mandatory presence of flow facts for the static WCET analysis, the lack of loop iteration counts prevents a compiler from applying various standard ACET optimizations. For example, without the knowledge about how often a loop iterates, WCC is not able to perform an effective loop unrolling [Muc97] or software pipelining [BGS94]. Thus, full optimization potential can not be exploited. A related domain of compiler optimizations which rely on the availability of loop iteration counts is *feedback-directed optimization* (*FDO*) [Smi00]. Exploiting this information enables a compiler to focus its optimization efforts on the most frequently executed program portions.

To relieve the compiler user from the burden of manual flow fact specification for a WCET analysis and to provide loop iteration counts for compiler optimizations, a static loop analysis for the ICD-C IR was developed in the course of this book.

The presented loop analysis is based on *abstract interpretation*, a theory of a sound approximation of program semantics. Abstract interpretation allows the computation of possible values that may be assigned to a program variable at a certain program point. Based on this information, the loop analyzer derives loop iteration counts for each loop in the program. To accelerate the analysis, the analyzed code is pre-processed using *program slicing*, a technique that excludes statements irrelevant for the loop analysis. Moreover, a novel *polyhedral loop evaluation* is introduced that further decreases the analysis time.

WCC's loop analyzer has proven to be of superior quality—among all tools participating in the WCET Tool Challenge 2008 [HGB+08], it was the only one which solved all flow facts related analysis problems.

The rest of this section is organized as follows. A survey of related work is given in Sect. 3.6.1, followed by an introduction to abstract interpretation in Sect. 3.6.2. The techniques of program slicing and the novel polyhedral evaluation are presented in Sects. 3.6.3 and 3.6.4, respectively. Finally, results achieved on real-life benchmarks are presented in Sect. 3.6.5.

3.6.1 Related Work

In [HSR+98], a pattern-based approach to determine loop iteration counts of assembly programs is presented. It exclusively evaluates instructions that represent loop code while other instructions are ignored. This way, loops relying on function parameters can not be analyzed. To overcome this problem, the authors provide a mechanism allowing to specify value ranges for unknown variables, making their analysis semi-automatic.

The approach developed in [HSR+98] has been adapted by Kirner [Kir06] to programs written in the high-level language C. Again, loop analysis does not automatically succeed for all types of loops. Mandatory information that can not be extracted during the static analysis must be provided by the user in the form of source code annotations.

In contrast to pattern-based loop analyses, an interprocedural data flow based loop analysis at assembly level is used in [CM07]. The advantage of this approach is that the success of the static analysis does not strictly rely on pre-defined code patterns generated by a particular compiler, but on the semantics of the instruction set for a specific target machine. As stated by the authors, their analysis works best for well-structured loops and supports only simple modifications of the loop counter.

A different approach for a fully automatic static loop analysis at source code level was described in [EG97]. The authors involve a data flow analysis which is based on *abstract interpretation*. The analysis computes information required to determine loop bounds. This work was used in [GES+06] to assist static WCET analysis. It was extended to support a determination of loop bounds for nested loops as well as a detection of *infeasible paths*, i.e., paths that are not taken in particular execution contexts of the program and which should thus be excluded from the WCET analysis to avoid WCET overestimation.

Further improvements to this loop analysis were presented in [ESG+07]. To accelerate loop analysis, the authors combine different standard program analyses like *program slicing* and *invariant analysis*. WCC's loop analysis additionally exploits slicing to enable a fast polyhedral loop evaluation.

3.6.2 Abstract Interpretation

The goal of a program analysis is to extract information about the program behavior. Such an analysis is based on the concrete semantics of a program. However, properties of concrete semantics of a computer system are in general undecidable as this computation is related to the *halting problem*. Another frequent problem of program analysis is its complexity, resulting from different states that must be considered (so-called *state explosion*). For example, a loop with n iterations containing a branch consists of 2^n possible paths each of which may be possibly considered as a different state.

Therefore, an approximation of program semantics has to be taken into account which guarantees termination and introduces a manageable computational complexity. Such an approximation should be sound, i.e., the computed properties are guaranteed to be true. A well-known theory of deriving such sound approximations is *abstract interpretation* [CC77].

To make WCC's loop analysis feasible for realistic applications, this book exploits abstract interpretation as the underlying framework for the computation of loop iteration counts. Using abstract interpretation, the complexity problems encountered during the computation of concrete semantics are tackled by replacing the concrete semantics domain by an abstract domain which require less computational effort. In addition, it is desirable to force a termination of WCC's loop analysis even if the program under analysis does not terminate. This goal is achieved by a safe extrapolation of the computed results during the analysis. For a detailed discussion about relevant aspects of abstract interpretation, the reader is referred to Appendix A.

3.6.2.1 Modified Abstract Interpretation

Abstract semantics is calculated by solving a set of data flow equations for which a least-fixed point is iteratively determined (cf. Appendix A). The iterative behavior of the classical abstract interpretation might significantly slow down the analysis such that it becomes impractical. In particular, such an explosion of analysis times can be observed for the analysis of loops with high iteration counts where each loop iteration is interpreted individually. The application of widening operators to accelerate abstract interpretation is not recommended as the operators introduce significant over-approximation which makes the results of loop analysis useless.

WCC's loop bound analyzer combines abstract interpretation with mechanisms to avoid its iterative behavior. These mechanisms rely on polyhedral loop analysis determining loop iteration counts and variable values by iterating the loop body exactly once. If these techniques succeed in computing loop bounds, classical abstract interpretation is omitted for a particular loop leading to an accelerated analysis. Otherwise, classical abstract interpretation needs to be applied.

The classical approach begins with the construction of a complete transition system consisting of data flow equations. Within this closed model, the solution is found by a fixed-point iteration. To avoid the iterative behavior of loops, a more flexible model of the transition system is required. WCC's loop analyzer does not consider the analyzed CFG as a closed monolithic system but subdivides the graph into smaller chunks for which a separate transition system is constructed and solved. This approach enables the application of different calculation methods to the individual transition systems. In particular, transition systems modeling loops can be computed by the fast polyhedral evaluation.

Like the classical abstract interpretation, WCC's loop analysis traverses the CFG starting at the program entry point. The abstract configuration assigned to this first program control point represents each variable either by $\top$ (denoting any possible value depending on the data type of the variable like `unsigned int`) or an interval that is defined by the user. The latter specification provides a way to communicate with the loop analysis in order to achieve more precise results that reflect a current environment.

Each node in the CFG represents an ANSI C statement and is associated with an abstract transition function which specifies the effect of this statement on the given abstract configuration. The transition functions are designed such that they compute a safe approximation of the statements' semantics on the interval domain. A comprehensive overview of the transition functions for available ANSI C statements can be found in [Cor08].

Example 3.4 The following figure gives an example for the abstract transition function of an **if-else**-statement based on the modified abstract interpretation.

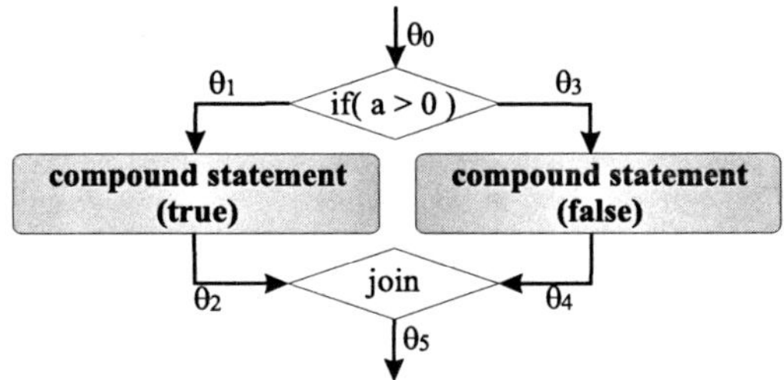

θ_n represents different program control points. The analysis starts with the abstract configuration $\hat{c}_{\theta_0}$. For configuration $\hat{c}_{\theta_1}$, all values from $\hat{c}_{\theta_0}$ which meet the condition **if(a > 0)** are assigned. Similarly, $\hat{c}_{\theta_3}$ is handled. These two abstract configurations are passed as input to the compound statement nodes representing the condition's **then**- and **else**-part, respectively. For these complex nodes, further transition functions are applied. Finally, their outputs represented by the abstract configurations $\hat{c}_{\theta_2}$ and $\hat{c}_{\theta_4}$ are merged at the *join* node leading to the computation of the abstract configuration $\hat{c}_{\theta_5}$.

If the fixed-point iteration is used as solver for the transition system of a loop, a counter variable is used in order to determine the number of loop iterations. The counter is incremented each time the analysis evaluates a transition function at the first control point within the loop body. If the CFG edge from the loop header node to the loop's immediate successor is traversed for the first time, the current value of the counter variable represents a safe lower bound for the number of loop iterations. The counter is incremented as long as the loop body is entered and a fixed-point for its abstract configurations is not achieved. If the fixed-point iteration has found a solution, the value of the counter variable represents a safe approximation for the upper bound of the loop iterations.

If the number of loop iterations depends on a variable represented by an interval with unequal bounds, e.g., generated by a *join* node, then the number of loop iterations might be variable in the sense that the lower and upper bounds are different. If a loop is executed in different contexts, its loop bound computation is performed multiple times and all context-sensitive results are collected. From these iteration counts, the minimal and maximal value represent the context-insensitive lower and upper bound loop bounds, respectively, that can be used as flow facts.

If the polyhedral evaluation is used for the analysis of an individual loop (as will be described in Sect. 3.6.4), the loop iteration counts can be statically computed in a single step.

To guarantee termination of WCC's loop analysis, the user can specify an upper bound ub which represents the maximal number of iterations each loop within the analyzed program may exhibit. During the analysis of each loop L, ub is taken into account. If this value is exceeded before the analysis could determine the loop iteration counts of L, the analysis of L is terminated and no results are available for this loop. Variables that are modified within L are, similar to the widening operator, extrapolated to their maximal value range depending on their data type. Afterwards, the loop analysis continues with subsequent nodes in the CFG following L. This way, the analysis tries to compute as many results as possible for loops in a given program.

3.6.3 Interprocedural Program Slicing

The static loop analysis operates on the ICD-C IR. For the acceleration of the loop analysis, a technique called *program slicing* [Wei79] is employed.

Definition 3.2 (Program slicing) *Program slicing* is a static program analysis which finds statements $s \in S$ that are relevant for a particular computation, where S represents all statements in the program $\mathcal{P}$.

The computation relies on the *slicing criterion* χ which is defined as:

Definition 3.3 (Slicing criterion) The *slicing criterion* χ is a pair $\langle \theta, \text{Var} \rangle$ where θ is a program control point and Var is a subset of program variables at θ.

Within WCC, slicing is applied on the ICD-C IR where $\theta \in S$ is represented by a statement. The result of program slicing is called a *slice* which is defined as:

Definition 3.4 (Slice) For a given program $\mathcal{P}$ and a slicing criterion $\chi = \langle \theta, \text{Var} \rangle$, the *slice* Γ defines a subset of $\mathcal{P}$ containing all statements s which may affect the variables in Var, i.e., variables that may either be used or defined at θ.

The fields of applications for slicing are manifold. Nowadays, program slicing is used for debugging [LW87, RR95], software maintenance [BH93], or compiler optimizations [FOW87].

An intermediate program representation that enables an efficient program slicing is the *program dependence graph (PDG)* [OO84]:

Definition 3.5 (Program dependence graph) A *program dependence graph* is a directed graph $G = (V, E, i)$, where nodes V correspond to statements. Edges $E \subseteq V \times V$ represent either control or data dependencies. i denotes a distinguished start node.

A *control dependence* is defined as follows:

Definition 3.6 (Control dependence) Let G be the control flow graph of a program $\mathcal{P}$. Let x and y be nodes in G. Node y is *control dependent* on node x if the following holds:

- there exists a directed path π from x to y
- y post-dominates every z in π (excluding x and y)
- y does not post-dominate x

Data dependencies represent essential data flow relationships of a program while the control dependencies, usually derived from a CFG, serve to indicate relevant control flow relationships. Taking both types of dependencies into account allows the efficient determination of a program slice with respect to a slicing criterion.

Example 3.5 The left-hand side of the following figure shows an example code and its PDG that consists of nodes representing statements and edges modeling data dependencies (dotted arrows) as well as control dependencies (solid arrows).

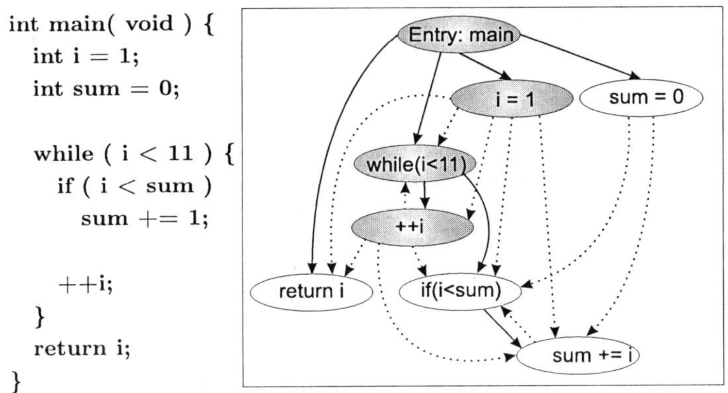

By definition, each PDG contains a distinguished entry node that serves as a starting point for the analysis. The program is sliced with respect to the **while**-loop exit condition **i** < **11**, i.e., **i** is a variable of the slicing criterion. The corresponding program slice can be found by reversing all edges and performing a *Depth First Search* algorithm starting at the node specified by the slicing criterion. All nodes that are visited are part of the slice. They are marked in gray in the shown figure and are those statements being relevant for the loop analysis. Statements in the remaining nodes can be omitted for the loop analysis.

The main drawback of the PDG is its restriction to an intraprocedural slicing. Since most real applications typically consist of numerous procedures, they can not be handled by PDGs. To overcome this problem, [HRB88] extended the PDG by concepts which enable to cross the boundaries of procedures. The interprocedural program representation is called *System Dependence Graph (SDG)*. It models program functions by PDGs and includes additional edges for the function calls as well as the transfer of values via function parameters and return values.

WCC's loop analysis applies program slicing on a SDG w.r.t. loop exit conditions. By taking all relevant data and control dependencies into account, the resulting program slice contains all statements that are involved in the determination of loop iteration counts. WCC's slicing is accompanied by a context-sensitive alias analysis to support pointers.

Slicing is run before the actual loop analysis for two purposes. First, in a similar fashion to [SEGL06], it aims at accelerating the loop analysis. By slicing the code in advance, all superfluous statements are stripped. Considering the relevant subset of the program, the fixed-point iteration during abstract interpretation can usually find a solution in shorter time. Second, the integration of the polyhedral loop evaluation, which is described in the next section, requires simple loop bodies to infer final abstract states without repetitive iterations. Loop bodies of original applications are often too complex for this static evaluation but after program slicing, the required prerequisites are frequently met.

3.6.4 Polyhedral Evaluation

A *polyhedron* P is an N-dimensional geometrical object defined as a set of linear equations and inequations. Formally, it is defined as follows [FM03]:

Definition 3.7 (Polyhedron & polytope) A set $P = \{x \in \mathbb{Z}^N \mid Ax = a, Bx \geq b\}$ is called a *polyhedron* for matrices $A, B \in \mathbb{Z}^{m \times N}$ and vectors $a, b \in \mathbb{Z}^m$ and $N, m \in \mathbb{N}$. A polyhedron is called a *polytope* if $|P| < \infty$.

Polytopes are often employed in compiler optimizations since they can be exploited to represent loop nests and affine condition expressions. Their formal definition enables efficient code transformations. Typical fields of application are program execution parallelization or the optimization of nested loops [FM03].

In this work, polytope models are applied for two purposes. On the one hand, they allow a precise computation of abstract configurations after processing a *condition* node which splits the control flow. On the other hand, for loops that are modeled by a polytope, its number of iteration counts can be determined statically. This knowledge is exploited by the non-iterative loop evaluation. The remainder of this section describes these two issues in more detail.

3.6.4.1 Polyhedral Condition Evaluation

The practical use of this approach will be motivated by an example. Assume that this condition is given:

$$\textit{if } (\overline{(2 * i + 2 \leq j \text{ NAND } j > 15)})$$

It can be represented by a conditional node as shown in Example 3.4. Furthermore, it should be assumed that abstract interpretation calculated the following abstract configuration before the evaluation of the condition:

$$\hat{c}_q = \{i \rightarrow [5, 15], j \rightarrow [10, 20]\}$$

The evaluation of the condition $2 * i + 2 \leq j$ yields for the configuration $\hat{c}_{q_{true}}$ and the values of $\hat{c}_q$ the condition $[12, 32] \overset{\wedge}{\leq} [10, 20]$. By definition, the result of

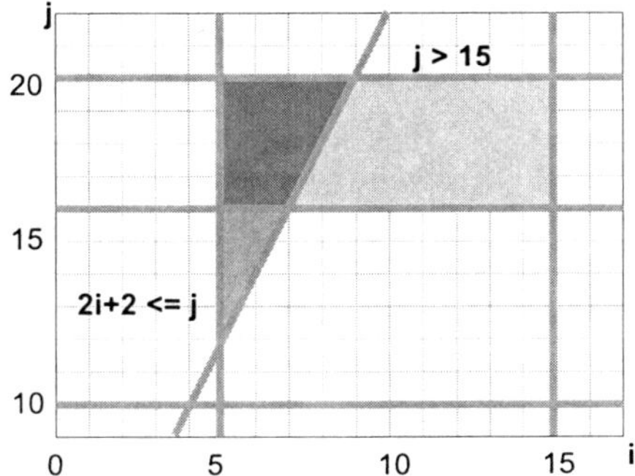

Fig. 3.6 Polyhedral condition evaluation

the abstract operator $\hat{\leq}$ is neither true nor false, thus it must be assumed that both outgoing edges of the condition node are taken. Due to the missing result of $\hat{\leq}$, the value ranges for i and j in the configurations $\hat{c}_{q_{true}}$ and $\hat{c}_{q_{false}}$ can not be precisely approximated. A conservative approximation for i and j in both configurations is to assume that their value ranges correspond to those computed for $\hat{c}_q$. Such a result is overestimated and often useless for further computations.

Since an effective determination of variable values for the successors of the condition is not feasible with the means of classical abstract interpretation, polytopes are used to determine precise values for the variables i and j. To compute polytopes, the condition must be transformed into the normalized form from Definition 3.7. See Appendix B for details of the transformation.

The finally transformed example

$$if\ (-2*i + j - 2 \geq 0 \wedge j - 16 \geq 0)$$

allows the construction of the two polytopes $-2*i + j - 2 \geq 0$ and $j - 16 \geq 0$ which are combined by the intersection operator $\cap$. These conditions (respecting the value ranges for i and j) are depicted in Fig. 3.6 as polytopes and their intersection is marked as dark area.

The final step is to approximate valid variable values for the abstract configuration that is valid if the condition is evaluated as *true*. The final polytope (intersection) encompasses all combinations of variable values that satisfy the condition. A safe approximation of values is defined by the extrema of the polytope, namely its corner points which are $(5, 16)$, $(7, 16)$, $(9, 20)$, $(5, 20)$. Performing a join over the coordinates, the abstract configuration $\hat{c}_{q_{true}}$ is defined as:

$$\hat{c}_{q_{true}} = \{i \rightarrow [5, 9], j \rightarrow [16, 20]\}$$

The configuration $\hat{c}_{q_{false}}$ is computed equivalently. Compared to the solution found by the classical abstract interpretation which defines the complete value range for i and j, this approach significantly increases the precision.

3.6.4.2 Polyhedral Loop Evaluation

A weak point of the classical abstract interpretation is its iterative behavior. To overcome this problem, the number of loop iterations is statically determined based on a polyhedral model. This knowledge is used to infer final values of variables modified in the loop without iterating over them repetitively.

Polyhedral loop evaluation within WCC is motivated by the observation that a large number of loops consists of statements not affecting the calculation of loop iterations. Typical examples are initialization procedures found in many embedded applications. The main task of such procedures is the initialization of arrays and other data structures. Afterwards, this initialized data is involved in the computation of output data, e.g., an output stream of an image compression algorithm, but it is not influencing the execution frequency of loops. Using program slicing, those statements are recognized to be meaningless for loop analysis and are removed for further evaluation. This frequently results in loops with almost empty loop bodies.

Loops to be analyzed by the polyhedral evaluation must meet certain constraints, specifying the structure of the loop and the type of statements in the loop body.

3.6.4.3 Preconditions for Loop Evaluation

The first class of requirements concerns the structure of loops including their conditional statements, e.g., `if`-statements. These restrictions are imposed by the polytope models and their violation would make a polytope evaluation infeasible, i.e., it would not be possible to determine how often a loop iterates and how often loop body statements within a condition are executed. The following preconditions must be met:

1. All loop bounds must be either constants or depend on a program variable which is not modified in the loop.
2. All `if`-statements must have the format `if` $(C_1 \oplus C_2 \oplus \cdots)$ where C_x are conditions that only depend on index variables of the loop nest and are combined with the logical operators $\oplus \in \{\&\&, ||\}$.
3. All conditions C_x depending on index variables must be affine expressions.

It should be noted that these constraints are often met by well-structured loops found in many embedded applications, thus they do not inhibit a successful application of WCC's polyhedral loop evaluation.

The second class of constraints refers to the loop body statements. If they are assignment statements, they must be transformable into the ANSI C assignment operators $=, +=$ or $-=$ to ensure that the variable is increased in each loop iteration by an additive value. Moreover, structs and pointers are not supported in the current version. Again, these requirements must be met to ensure that a polyhedral evaluation can be performed. Even though these constraints seem to be highly restrictive, experiments have shown that sliced loop bodies often satisfy them. For a detailed description of the preconditions, the interested reader is referred to [Cor08].

3.6.4.4 Ehrhart Polynomial Evaluation

If the conditions are met, the loop iteration counts required for the evaluation of statements are statically determined in the next step. Results from this phase allow a fast static evaluation of statements in a single step without the need to analyze the statements iteratively. The problem of finding the loop iteration counts is equivalent to computing the number of integer points in a polytope. To efficiently count the integer points, *Ehrhart* polynomials [VSB+04] are used. Techniques for transforming polytopes that represent a loop nest into an Ehrhart polynomial are omitted here since they go beyond the scope of the book.

Taking all integer points of a polytope might yield an over-approximation. The total number of integer points represents the number of loop iterations if the loop counter is incremented by one, thus other modifications to the counter must be adequately modeled. Also, additional exit edges that affect the control flow in the loop body, e.g., in the case of `break`- or `continue`-statements, must be taken into account. They are modeled as further polytopes and their intersection with the polytope representing the loop nest yields the precise solution space. For some loops found in real-life benchmarks having an empty loop body after program slicing, counting of integer points is already sufficient to determine the loop iteration counts statically.

Using these loop iteration counts, execution frequencies of condition-dependent basic blocks, which might obviously differ from the loop iteration counts, are computed. The conditions are modeled by polytopes and an intersection with the loop polytope allows to compute execution frequencies for both the `then`- and `else`-part.

Example 3.6 An example for the determination of iteration counts for a nested loop is depicted in the following figure.

```
for ( i=1; i<= 8; ++i ) {
    for ( j=i; j<=15; ++j ) {
        if ( i > 2 ) {
            ...
        }
    }
}
```

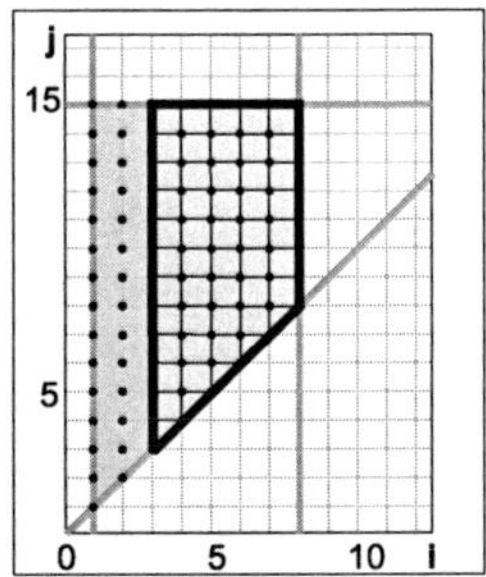

The one-dimensional polytope representing the outer loop holds 8 points. The iteration space for the inner loop depends on the outer loop and its corresponding polytope is indicated by the gray area. Using Ehrhart polynomials, the number of integer points for this polytope can be computed and equals 92. The shaded polytope represents the **then**-part of the **if**-statement in the inner loop nest. From the number of integer points it can be inferred that this code fragment is executed 63 times.

3.6.4.5 Static Statement Evaluation

The last step is the static evaluation of statements within the loop based on the loop iteration counts and basic block execution frequencies from the previous step. The goal is to evaluate modifications of variables within the loop like assignments b+=a without a repetitive abstract interpretation. These final variable values are used to determine iteration counts for loops that are analyzed afterwards.

Example 3.7 Assume that within the **if**-statement of the code in Example 3.6, the assignment $a+ = 2$ is executed. If the abstract configuration of **a** before the execution of the loop nest is known, the final value of **a** can be statically inferred. For example, if $\{a \to [0, 5]\}$, the finally value of a is $\{a \to [126, 131]\}$ since this statement is executed 63 times.

WCC's loop analysis can be used in two different scenarios. It can either be used as a stand-alone tool producing loop information in a human-readable form, or as a module integrated into WCC to assist optimizations or static WCET analyses. In the latter case, the loop analyzer automatically generates flow facts and passes them to the compiler relieving the user from manually annotating flow facts. Figure 3.1 on p. 27 depicts the analyzer's integration into WCC. It operates on the ICD-C source code level and starts with program slicing marking statements relevant for loop bound computations. After that, loop analysis based on the modified abstract interpretation and polytope models using the library *Barvinok* [Ver09] is performed. At this point, the loop bound information for the program under analysis can be generated.

3.6.5 Experimental Results

To indicate the efficacy of WCC's loop analyzer, evaluations on a large number of benchmarks were conducted. The benchmarks come from the test suites DSPstone [ZVS+94], MRTC [MWRG10], MediaBench [LPMS97], MiBench [GRE+01], UTDSP [UTD10], and WCC's collection of real-life benchmarks containing miscellaneous applications, e.g., an H263 coder or a G.721 encoder. The different types of the suites were chosen to point out that WCC's loop analysis can successfully handle applications of different domains.

The results show how many loops could be precisely analyzed. All measurements were performed on a single core of an Intel Xeon CPU with 2.40 GHz and 8 GB RAM. In total, 96 benchmarks were extensively analyzed and evaluated. For the sake of clarity, a comprehensive overview of the results is provided and the interesting cases are discussed in more detail in the following.

Table 3.3 Precision of loop analysis

Benchmark Suite	# Benchmarks	# Loops	Analyzable	Exact
MRTC	32	152	100%	99%
DSPStone	37	152	98%	93%
MediaBench/MiBench	6	162	99%	98%
UTDSP	14	88	100%	88%
Misc.	7	153	100%	100%
Total/Average	96	707	99%	96%

3.6.5.1 Determination of Loop Iteration Counts

Table 3.3 presents the evaluation of the loop analysis precision. The table shows for each benchmark suite the number of benchmarks, the number of contained loops, the percentage of loops that were successfully analyzed (column *Analyzable*) and the percentage of loops for which the loop analysis produces exact non-over-approximated results (column *Exact*). All percentages of Table 3.3 relate to column *Loops*.

The 96 benchmarks contain 707 loops in total. On average, 99% of those loops could be successfully analyzed. This means that for those loops, loop analysis produced safe results in terms of loop iteration counts which are never under-approximated, but might be over-approximated. The small fraction of 1% loops that could not be analyzed is mainly due to technical restrictions of WCC's alias analysis.

The last column of Table 3.3 presents the ratio of loops that are exactly analyzable, i.e., loops for which precise loop iteration counts were determined. On average, the analysis produced exact results for 96% of the loops. The remaining 4% of the 707 loops, including the non-analyzable 1% of loops from the previous column could not be exactly analyzed, i.e., the loop iteration counts were afflicted with an over-approximation. The main reason for the imprecision comes from the analysis of pointers which can not always be precisely evaluated in a static analysis. However, most of the over-approximations introduced only a marginal error ranging between 8% and 51% w.r.t. the exact results. Thus, the results are still acceptable.

Program slicing was successfully applied to all benchmarks. The number of statements irrelevant for the loop analysis ranges between 2% and 88% showing that computations in many programs do not affect the loop iteration counts. 21% of the 707 loops could be analyzed with the polyhedral loop evaluation. This number indicates that the constraints for a successful application of this evaluation are not too restrictive and are often met in real-life applications.

3.6.5.2 Analysis Time

Besides the precision of the analysis, the second crucial issue for static program analyses is their complexity expressed in terms of analysis time. In general, the analysis times highly depend on the program structure and the loop iteration counts. If

Table 3.4 Run times of loop analysis

Benchmark	Benchmark Suite	Basic	Slicing	Polytope
matmul	MRTC	8.4 s	2.4 s (28%)	0.8 s (1%)
hamming	Misc.	0.4 s	0.3 s (80%)	0.2 s (62%)
g721	DSPstone	80.2 s	70.5 s (88%)	71.3 s (89%)
fft	DSPstone	920.7 s	119.7 s (13%)	110.5 s (12%)
matrix1	DSPstone	0.8 s	0.09 s (12%)	0.03 s (4%)
mult_10_10	UTDSP	4.6 s	3.6 s (78%)	3.7 s (80%)

the polyhedral loop evaluation can not be applied, the analysis based on abstract interpretation must consider each loop iteration separately. On average, smaller benchmarks require a few seconds for the analysis, while the analysis time for larger benchmarks such as MiBench's `gsm_encoder` takes on average less than 4 minutes.

The impact of the different techniques on the analysis run time of some example benchmarks is shown in Table 3.4. Column *Basic* represents the absolute run time of the basic loop analysis based on abstract interpretation. The fourth column (*Slicing*) depicts the analysis run time after program slicing, while the last column (*Polytope*) indicates the measured run times after the application of the fast polyhedral loop evaluation (including slicing). In addition, values in parentheses found in the fourth and fifth column represent the relative run times w.r.t. the third column.

Table 3.4 shows that slicing significantly decreases analysis times. For *matmul*, a reduction of 72% was achieved. *matmul* also benefits from the polytope approach. It contains some loops that can be statically evaluated using the polyhedral model, leading to a further reduction in time of 27%. For other benchmarks, like *mult_10_10*, slicing reduces the analysis time by 22%. For *mult_10_10* the test whether the polytope approach can be applied was negative, thus slightly increasing the analysis time by 2% and forcing the analysis to switch back to the basic (iterative) loop evaluation.

Considering all 96 evaluated benchmarks, 38 benchmarks benefit from program slicing leading to a decreased analysis time. For 13 of these benchmarks, the analysis time could be further improved by switching from the iterative approach based on abstract interpretation to the polyhedral approach.

The approach presented in this section was published in [LCFM09].

3.7 Back-Annotation

The integration of a WCET analyzer into WCC's compiler backend enables the exploitation of the worst-case timing data by assembly level optimizations. However, to exploit the high potential of source code level optimizations for WCET minimization, the timing data must be translated from LLIR into ICD-C IR. The connection of different abstraction levels of the code and an exchange of data between these levels is carried out in the WCC framework through a process called *back-annotation*. In the following, the concepts and implementation of back-annotation developed in this work are presented.

Problem 3.2 (Back-annotation) Given two code representations $\mathcal{P}_{high}$ and $\mathcal{P}_{low}$, with $\mathcal{P}_{high} \stackrel{.}{>} \mathcal{P}_{low}$, where the operator $\stackrel{.}{>}$ expresses the relation between the abstraction levels of the code representations, i.e., $A \stackrel{.}{>} B$ indicates that code representation A contains less information than B w.r.t. the execution of program $\mathcal{P}$ on a given target architecture. The problem of the back-annotation is (a) to find a unique mapping from components of $\mathcal{P}_{low}$ to equivalent components of $\mathcal{P}_{high}$ such that $\text{backan}_{\mathcal{P}} : \mathcal{P}_{low} \rightarrow \mathcal{P}_{high}$, and (b) to exploit this mapping for an information exchange between the lower- and higher-level code representation.

3.7.1 Mapping of Low-Level to High-Level Components

The connection between the LLIR and the ICD-C IR is established during code selection. This is the only phase where the compiler has the full knowledge about structural relationships between the low- and high-level components. To realize the mapping *backan*, the back-annotation uses unique identifiers that map an LLIR component to a corresponding ICD-C component. The type of the identifiers depends on the level of granularity on which the back-annotation is performed.

At the most coarse-grained level, mapping of compilation units (source code files) using the file name as identifier is trivial. During code selection each ICD-C compilation unit with a unique name is translated into a single LLIR compilation unit. In a similar way, each high-level function is translated into one low-level function. Using the unique function name as identifier yields a bijective mapping at this level of granularity. An exception are ANSI C *static* functions [Sch00] which may have identical function names in different compilation units. To achieve unambiguous identification, the function names used for static functions are concatenated with the compilation unit name.

The finest granularity level exploited for WCC's back-annotation are basic blocks using the block mapping function defined as

$$\text{backan}_b : b_{low} \rightarrow b_{high}$$

However, mapping of LLIR and ICD-C blocks is not trivial due to a missing bijection. The reasons for the encountered $n{:}m$ mapping are twofold. First, the definition of basic blocks deviates in both IRs w.r.t. handling of calls. Calls constitute a block's boundary in the LLIR but not in the ICD-C IR. Second, some implementation-specific issues complicate the block mapping. For the sake of completeness, possible scenarios and solutions to them are briefly discussed in the following:

- **1:1 Relation**

    Sequential code without control flow modifications is represented in the LLIR and ICD-C IR as a single block. This bijection is exploited and a mapping between the blocks is achieved using the block label as unique identifier.

- **n:1 Relation**

    Due to different definitions of basic blocks, a source code like

```
{
    a = a / 2;
    a + = 100;
    a = foo( a );
    return a;
}
```

is represented by two LLIR blocks but a single ICD-C block. Another reason for this scenario are the logical ANSI C operators AND (&&), OR (| |), and the conditional operator (?) which implicitly modify the control flow. At assembly code level, they are modeled by jump instructions resulting in multiple basic blocks. However, at source code level a statement constitutes the modular unit of a basic block, thus block boundaries can not be modeled within a statement. Thus, statements containing these operators are covered by a single ICD-C block. As a consequence, block mapping might be surjective with several LLIR blocks mapped to a single ICD-C block.

- **0:1 Relation**

    Assembly level transformations may delete LLIR basic blocks, e.g., a low-level dead code elimination. Since these modifications to the set of LLIR blocks are performed after code selection, they do not affect ICD-C blocks. However, to have a consistent mapping between both IRs, such transformations must notify the back-annotation which updates the mapping data.

- **1:m Relation**

    For technical reasons, the header of an ICD-C loop is always modeled by a separate basic block. Simple loops such as

```
do {
    a - = 1;
    sum + = a;
} while ( a > 0 );
```

may be differently covered by WCC's IRs. Within ICD-C IR, the loop body and the loop header are represented both by a separate basic block. In contrast, this loop is modeled by a single LLIR basic block since it does not contain any control flow modifications. For such 1:m relations, the LLIR block must not be mapped to all m ICD-C blocks since this mapping would equal an undesired duplication of data. For example, the imported WCET data for the do-while-loop from the previous example would be twice as large as estimated by aiT, leading to a falsified worst-case timing model. Thus, timing information is attached to one of the *m* ICD-C blocks which is called the *reference block*. Any requests to the remaining $m-1$ ICD-C blocks are forwarded to the respective reference block in order to access the desired back-annotation information.

3.7.2 Back-Annotated Data

Back-annotation provides a general framework for the exchange of arbitrary information between a low-level and high-level IR. Primarily, WCET data and the worst-case execution counts (*WCEC*) determined by aiT are imported into the ICD-C IR. But also other information is exchanged. In detail, WCC's back-annotation translates the following data:

- WCET for each basic block, function, compilation unit, and entire program
- Information whether a block lies on the WCEP
- Execution feasibility of blocks and CFG edges
- WCEC per block and CFG edge
- Worst-case call frequencies between functions
- Number of I-cache misses per block encountered during WCET analysis
- Low-level block size in bytes
- Number of spill code instructions per block as generated during register allocation

Due to a missing support of contexts in the LLIR, the WCET-related data listed above represents values accumulated over all execution contexts. The transformation of data during back-annotation traverses the LLIR CFG and assigns each ICD-C block b_{high} information annotated at the given LLIR block b_{low} exploiting $backan_b : b_{low} \mapsto b_{high}$. For the n:1 block relation, data from the n LLIR blocks is merged according to the following rules (merging for CFG edges works analogously):

- Accumulate WCET information: $WCET(b_{high}) = \sum_{k=1}^{n} WCET(b_{low}^k)$
- Assign maximal WCEC: $WCEC(b_{high}) = \max\{WCEC(b_{low}^k) \mid \forall k = 1, \ldots, n \}$
- If at least one WCEP attribute of the n blocks b_{low} is set to *true*, set $WCEP(b_{high}) = true$, otherwise $WCEP(b_{high}) = false$
- If at least one *infeasible* attribute of the n blocks b_{low} is set to *false*, set $inf(b_{high}) = false$, otherwise $inf(b_{high}) = true$

In typical embedded applications, the WCEC, WCEP, and *infeasible* attributes are equal for all n LLIR blocks.

3.8 TriCore Processor

This chapter concludes with a brief description of the target architecture used in the WCC compiler framework. Since many of the compiler optimizations discussed in

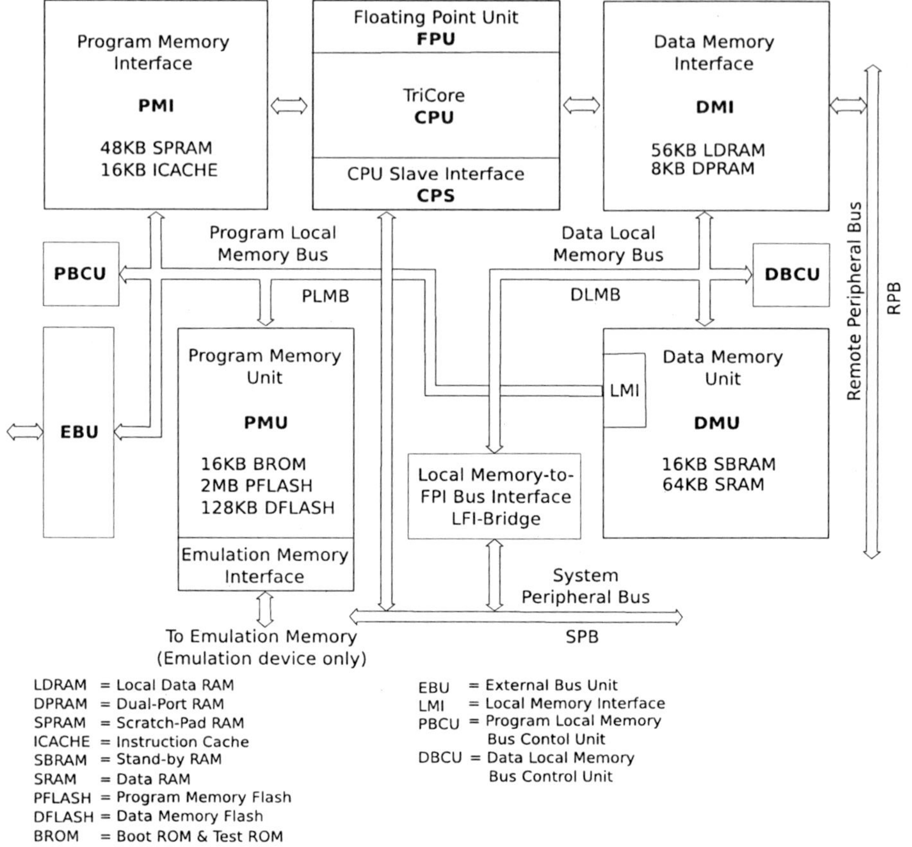

Fig. 3.7 TriCore TC1796 architecture

the following chapters exploit specific features of the underlying processor, the most relevant components of the processor will be summarized in the following.

The current implementation of WCC targets the *TriCore TC1796* V1.3 processor [Inf08a] developed by *Infineon Technologies*. It is a single-core 32 bit DSP architecture optimized for embedded systems. The processor's frequent utilization in safety-critical automotive applications and its sophisticated architecture providing high potential for compiler optimizations were the major reasons for its selection as WCC's target processor.

TC1796 is a RISC 32 bit load/store Harvard architecture running maximally with 150 MHz. The TriCore instruction set architecture consists of 16 and 32 bit instructions for high code density. In addition to general-purpose instructions, numerous DSP-specific instructions, such as multiply-accumulate (MAC), SIMD, or saturation, are available. The processor provides fast hardware controlled context switches taking 2–4 cycles and a support for a floating point unit.

The processor is equipped with a 4-stage pipeline using a triple issue super-scalar implementation allowing a simultaneous execution of up to 3 instructions, called *bundles*. Most instructions are executed either on the *integer* or *load-store* pipeline, while the *loop* pipeline is dedicated to the execution of zero-overhead loop instructions.

The register file consists of 32 general purpose registers (GPR) that are divided into 16 data and 16 address registers. Moreover, two of the 32 bit registers can be accessed as an extended 64-bit register.

Memory hierarchies play an eminent role in compiler optimizations for embedded systems. An overview of the memory and bus systems of the TC1796 [Inf08a] is depicted in Fig. 3.7. Next to the CPU, L1 memories, called *scratchpads*, for program code (PMI) and data (DMI) are located. In addition to these compiler-controlled memories, the TC1796 is equipped with a (hardware-controlled) instruction cache. Further memories are the L2 data memory (DMU) and a program and data Flash. Additional external memories can be integrated via the external bus unit (EBU).

Besides the TriCore as the core architecture, a first approach to extend the WCC framework towards a multi-target system was presented in [PLM08]. Using this methodology, an integration of the ARM7TDMI processor [ARM01] into WCC is in progress.

Chapter 4
WCET-Aware Source Code Level Optimizations

Contents

P. Lokuciejewski, P. Marwedel, *Worst-Case Execution Time Aware*   61
Compilation Techniques for Real-Time Systems, Embedded Systems,
DOI 10.1007/978-90-481-9929-7_4, © Springer Science+Business Media B.V. 2011

4.1 Introduction

High-performance embedded systems can only be developed when efficiency requirements are pursued at different levels of the system design. A predominant role is associated with compilers which are responsible for the generation of efficient machine code. To accomplish this goal, compilers have to feature advanced optimizations. The class of source code optimizations provides a number of benefits compared to optimizations applied at lower abstraction levels of the code. The most important issues are briefly discussed in the following.

First, compiler optimizations applied at source code level are characterized by their portability. Any optimization developed for a particular high-level programming language can be reused in the compiler without any modifications after the exchange of the target architecture in the compiler backend. Regarding WCET-aware compilation, portability is also preserved. A prerequisite for WCET-aware source code level optimizations is a back-annotation (cf. Sect. 3.7) which transforms WCET data from the low-level IR into the high-level IR. This data is transparent to the compiler frontend and can be employed for source code optimizations without further consideration of the underlying hardware. This mechanism makes a compiler more flexible and fosters reusability of software, leading to a significantly decreased development time of WCET-optimizing compilers.

Second, the application of source code level optimizations establishes opportunities for subsequent code transformations. Assembly level optimizations often benefit from code modifications performed in the compiler frontend. For example, high-level loop unrolling has shown to have a positive effect on low-level instruction scheduling since an unrolled loop body increases instruction level parallelism. But also source code optimizations may benefit from each other. A well-known example is function inlining. By replacing the function call by the body of the called function, additional potential for intraprocedural optimizations, such as local common subexpression elimination, is enabled that would not have been possible without inlining.

Finally, the high level of abstraction turns out to be advantageous for source code optimizations. A high-level representation of the code provides more details about the program under analysis and this knowledge can be exploited for a sophisticated code transformation. To illustrate this issue, consider the transformations of an ANSI C code snippet performed by loop deindexing (also known as *lowering array-based code to pointer-based code* [LPJ96]):

```
for (int i=0; i<a; ++i) {          int *_1=A, *_2=B; {
  A[i]=B[i];              →          for (int i=0; i<a; ++i)
}                                       *_1++=*_2++;
                                    }
```

As can be seen on the right-hand side of the example, the access to arrays A and B has been replaced by pointers. The optimized code allows a translation into more efficient assembly code since the base address of the arrays is loaded before loop execution and the pointers can be covered by cheap auto-increment load/store instructions. In contrast, loop deindexing is not applicable at assembly level due to missing information about data types such as ANSI C arrays or structs. In addition, source code optimizations can benefit from static program analyses that are only available at the high abstraction levels of the code. Examples for such techniques include alias analysis [Muc97] for the evaluation of pointer expressions and advanced static loop analysis (cf. Sect. 3.6) which can not be efficiently applied to assembly code.

In this chapter, WCET-aware source code level optimizations applied in WCC's frontend ICD-C are presented. All of the presented techniques rely on a back-annotation which provides WCET information in ICD-C. Section 4.2 gives a survey of work related to WCET-aware source code level optimizations. In Sect. 4.3, the optimization procedure cloning is presented. It will be shown that the influence of cloning may significantly differ on the ACET and WCET of the program. This knowledge is exploited for a WCET-aware procedure cloning which focuses on an effective WCET minimization while keeping the inherent code expansion small. In Sect. 4.4, a structure called superblock is constructed by means of WCEP information and exploited for WCET minimization. A WCET-aware loop unrolling that makes extensive use of results from WCC's static loop analysis as well as information about the memory hierarchy will be described in Sect. 4.5. In order to accelerate optimizations prone to a WCEP switch, Sect. 4.6 introduces the concept of the invariant path and demonstrates its effectiveness for the optimization loop unswitching. Finally, a short summary in Sect. 4.7 concludes this chapter.

4.2 Existing Code Optimization Techniques

Source code optimizations are popular compiler techniques for the optimization of the average-case performance. Most standard optimizations improve the code quality without prioritizing any code sections, i.e., each CFG path is treated equally. They include for instance value propagation, dead code elimination, or loop collapsing [BGS94]. Another class of optimizations is profiling-based optimization. In order to identify hot spots in the code, i.e., regions in the code that considerably contribute to the total execution time of the program, numerous compiler frameworks interact with profilers. Profiling is typically done at the assembly/machine code level demanding a back-annotation for the import of execution frequencies into the compiler frontend. However, an advanced back-annotation, which achieves a proper mapping between the original source code and the profiled optimized machine code, is often not available in many compilers. Due to this reason (aka. *code location problem* [VRF+08]), the compiler user is often forced to gather profiling data for non-optimized code which is an undesired restriction.

WCC's back-annotation does not suffer from the code location problem. The mapping between the source code and the assembly code is established in the code selector after ICD-C optimizations were performed. Moreover, any LLIR optimizations applied in the compiler backend automatically update the mapping information (cf. Sect. 3.7). Therefore, a correct connection between both abstraction levels is given at any time.

Unlike ACET optimizations, WCET-aware source code optimizations are hardly available in today's compilers. The reasons are twofold. First, a WCET minimization relies on an advanced compiler infrastructure that communicates with a WCET analyzer. Since WCET-aware compilation is still a novel research area, the number of qualified compilers is marginal as indicated in Sect. 3.2. The few published works restrict themselves to the domain of low-level optimizations since the compiler backend is the location where the WCET analysis takes place, avoiding the need for a back-annotation. A detailed overview of these optimizations follows in Chap. 5.

The mandatory existence of a back-annotation is the second reason for the lack of WCET-aware source code optimizations. To the best knowledge of the author, besides WCC the *TUBOUND* compiler framework [PSK08] is the only functional compiler which supports a back-annotation. However, no source code optimizations were reported yet.

Besides the approaches developed in this book, two related works concerning a compiler-based WCET minimization at source code level have been published. In [FS06], the impact of an ACET optimization on the worst-case performance was evaluated. The authors applied the optimization loop nest splitting in a standard compiler to increase the predictability of the code. The optimization transforms loop nests at source code level such that complex `if-then-else`-statements are simplified or even eliminated, leading to a more homogeneous control flow structure. The static WCET analysis benefits from these code modifications since overestimations emerging during the pipeline and branch prediction analysis of jump instructions are diminished. In addition, the transformed code allows a more accurate specification of flow facts for loops. An evaluation of loop nest splitting on real-life multimedia applications indicated that the average WCET on an ARM7 processor was reduced by 36.3%, while the average ACET reduction amounts only to 30.1%.

A compiler-guided trade-off between WCET and code size for an ARM7 processor was studied by [LLPM04]. The authors observed that applications implemented with 16 bit THUMB instructions are smaller but also slower than the same code using the full 32 bit instruction set. They used a simplified timing analyzer to obtain WCET information. This data was employed in their code generator to produce code that is aware of this trade-off and uses the two instruction sets for different program fragments.

WCC is currently the only compiler which supports WCET-aware high-level optimizations. An overview of the currently available techniques is presented in the following section.

4.3 Procedure Cloning

Procedure cloning, also known as function specialization, is a well-known compiler optimization typically applied at source code level. Its goal is the improvement of the ACET by a reduced function call overhead. Moreover, it is an enabling transformation for other optimizations, i.e., a prior cloning enables the application of optimizations that would not have been possible without the cloned code. In this book, procedure cloning is studied in the context of the WCET.

The remainder of this section is organized as follows. Section 4.3.1 discusses problems emerging during a static WCET analysis of embedded real-time applications and proposes procedure cloning as a possible solution. An overview of related work to cloning is presented in Sect. 4.3.2, followed by a description of the standard cloning algorithm in Sect. 4.3.3. In Sect. 4.3.4, the impact of procedure cloning is demonstrated on real-life applications. These results indicate that the improvements come at the cost of a substantial code growth. Therefore, a WCET-aware procedure cloning, which exploits the benefits of the optimization but significantly limits the adverse code expansion, is proposed in Sect. 4.3.6. The experimental results of this optimization are presented in Sect. 4.3.7.

4.3.1 Motivating Example

Static WCET analysis relies on flow facts (cf. Sect. 3.5) which specify loop iteration counts and recursion depths. Since typical embedded applications spend most of their execution time in loops, flow facts have a significant impact on the application's WCET and their precise specification is mandatory for tight WCET estimations. However, code structures found in typical applications complicate a precise loop analysis.

Many embedded real-time applications are data-dominated. Their source code often contains data-dependent loops:

Definition 4.1 (Data-dependent loop) Let $\mathcal{L}$ and $\mathcal{F}$ be the set of loops and functions of a program $\mathcal{P}$, respectively. A loop $L \in \mathcal{L}$ is called a *data-dependent loop* if its loop iteration counts depend on a variable *par* which is a parameter of function $F \in \mathcal{F}$ and $L \in F$. Parameter *par* affects the loop exit condition which is either placed in the loop header or the loop body.

A review of these data-dependent applications revealed that functions F containing data-dependent loops are often called from various locations within $\mathcal{P}$. In addition, the calls use different values for the arguments translated to the parameters *par* of F. Hence, the effective number of loop iterations within F can vary considerably, depending on the execution context, i.e., how and with which parameters F is called.

Example 4.1 Code from the *gsm* benchmark found in the MediaBench Suite [LPMS97] illustrates this situation:

```
void Gsm_Short_Term_Analysis_Filter (...) {
   ...
   Short_term_analysis_filtering(S, LARp, 13, s);
   ...
   Short_term_analysis_filtering(S, LARp, 14, s);
   ...
   Short_term_analysis_filtering(S, LARp, 13, s);
   ...
   Short_term_analysis_filtering(S, LARp, 120, s);
   ...
}

static void Short_term_analysis_filtering (struct gsm_state * S,
          word * rp, int k_n, word * s) {
   ...
   for (j = 0; j < k_n; ++j) {
      ...
   }
}
```

The filter function **Short_term_analysis_filtering** computes the short term residual signal. It is called four times in **Gsm_Short_Term_Analysis_Filter** with constant arguments varying between **13** and **120**. These arguments are translated in
Short_term_analysis_filtering into the parameter **k_n** that controls the number of
loop executions of the **for**-loop.

To enable a precise static program analysis, different calls to functions with data-
dependent loops must be distinguished in order to respect the varying loop behavior. Modern WCET analyzers like aiT support the concept of execution contexts
to represent the history of the invocation of a function. By distinguishing between
different contexts, the variable values that are valid at a certain function call can be
determined. That way, each function call is analyzed separately with precise parameter values. The context information is propagated to the code in the function body
enabling a tight analysis of data-dependent loops for which precise loop iteration
counts are regarded.

However, the support of context-sensitivity enhances not only the analysis precision but also significantly increases its complexity [The02]. In particular, nested
loops can cause a rapid increase of contexts leading to *state explosions*. For example, the loop nest of three nested loops with ten iterations each requires 1000
contexts to be distinguished. Such large numbers of contexts can make a context-
sensitive WCET analysis infeasible since both a vast amount of memory resources
and computation time are necessary.

This phenomenon is depicted in Fig. 4.1. For three complex real-life applications,
the figure shows the time taken for WCET analysis using aiT for different (small)
numbers of considered execution contexts. As can be seen, WCET analyses require
exponential run times for increasing numbers of contexts. For example, the analysis

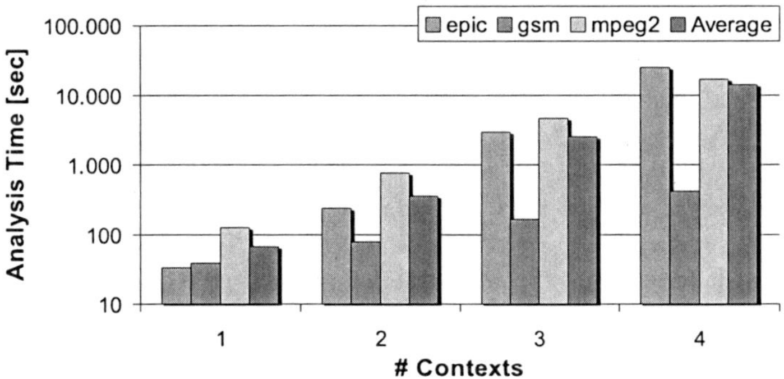

Fig. 4.1 Context-sensitive WCET analysis times

time for the *mpeg2* benchmark took 125 seconds for one and 16,723 seconds for four distinguished contexts.

In order to decrease the analysis time and to make the WCET analysis feasible, the number of distinguished contexts can be restricted. However, this loss of information has a negative effect on the precision of the timing analysis of loops. Restricting the number of contexts to n means that the first $n - 1$ loop iterations are represented by a distinguished context, leading to a precise analysis. If the number of loop iterations u is larger than n, the remaining $u - n + 1$ iterations are summarized by the last context. In order to obtain a safe timing estimation, the analyzer determines the maximal WCET per iteration among the remaining iterations. This WCET has to be pessimistically assumed for each of the $u - n + 1$ iterations, possibly yielding a high overestimation.

But not only the context restriction leads to imprecision during loop analysis. The structure of loops in real-life applications is often too complex to be analyzed at assembly level. Therefore, aiT's loop analysis only succeeds for a limited class of loops. Due to these reasons, flow facts for most loops require either a manual specification by the user or by WCC's static loop analysis. The common form of flow facts for loops is a $[min, max]$ interval (cf. Sect. 3.5) for each loop, modeling lower and upper bounds of the possible number of iterations of a loop.

The specification of flow facts for data-dependent loops turned out to suffer from the interval representation. Since function F containing such loops is called from many places with possibly different arguments, the effective number of loop iterations within F can vary considerably, depending on how and with which parameters F is called. However, the flow facts for such a loop must cover all these different contexts in which the loop may be executed in order to result in safe WCET estimation. Hence, the lower bound of such flow facts must represent the global minimum of iterations executed by such a loop over all contexts in which F is called, and the same holds for the upper bound. Since such flow facts for data-dependent loops do not consider possible different execution contexts of a function F, the flow facts are safe but lead to a highly overestimated WCET.

Procedure cloning (also known as *function specialization*) is a standard optimization [Muc97] exploiting functions that are often called with constant values as arguments. If the caller invokes a callee with some constant parameters, the callee can be cloned by removing the constant parameters from the list of parameters and importing them into the callee itself. In standard literature, the benefits of cloning are said to be twofold. First, procedure cloning potentially enables further optimizations, e.g., constant propagation and constant folding within the clone. Second, the calling overhead is reduced since the constant parameters are not passed any more between caller and callee.

In this book, the impact of procedure cloning on the WCETs of embedded real-time applications is investigated and it is shown that cloning can be highly effective for the WCET analysis. The transformation creates a specialized version $\hat{F}$ of F. If F contains data-dependent loops, the data-dependence of the number of loop iterations is resolved since constant argument values are propagated into the loop exit conditions. The optimized code becomes more predictable since precise flow facts for the data-dependent loops in $\hat{F}$ can be provided. Hence, procedure cloning can be seen as an approach to make different calling contexts explicit at the source code level, thus enabling a high-precision WCET analysis of these specialized functions. An important advantage of this technique is that the WCET reduction is achieved without noticeably increasing analysis times compared to a context-sensitive analysis as indicated in Fig. 4.1.

In addition, the tightness of the WCET estimates can be further improved by the elimination of infeasible paths. Such paths are executable according to the control flow graph but are not feasible when taking the program semantics and possible input values into account. The propagation of constant function arguments into the function bodies helps to identify infeasible paths in the function clones, thus improving the static WCET analysis.

4.3.2 Related Work

Procedure cloning was initially published by Cooper in [CHK93] and is now part of standard literature on compiler construction [Muc97]. Up to now, cloning was exclusively studied in the context of ACET and code size.

Cooper addressed the problem of interprocedural transformations and exploited cloning to enable compiler optimizations that could previously only be performed after function inlining which often entails an undesired code expansion. Before cloning a function, the compiler evaluates the potential for further optimizations that would arise if a particular function was cloned. This way, only promising function candidates are transformed, resulting in high average-case performance improvements and marginal code size increases.

A similar approach was pursued in [SES00]. The authors observed that Fortran 90 subroutine calls can cause serious performance losses if copy-in/copy-out argument passing is required. To prevent redundant argument copying, a cloning algorithm utilizing data-flow analyses was presented. It enables the identification of profitable procedure clones and a controlled code growth. The approach presented in [SD03] explored the impact of procedure cloning on the instruction level parallelism of VLIW and EPIC machines. By merging function clones, multiple independent instruction streams were generated which can be executed in parallel.

In [Vah99], procedure cloning was studied in the context of functional partitioning. The authors applied the transformation to create function clones that can be assigned to different processors which execute the calls to these clones. The result is a performance improvement due to the reduced communication overhead.

A related work to procedure cloning was explored in the field of feedback-directed optimization [FMP+07]. Motivated by the fact that the behavior of a program can vary considerably across different data sets, the authors cloned the most time-consuming functions during compilation and applied different standard optimizations to them. In a second step, the program was run with different inputs on different target architectures and the performance of the original and cloned functions was evaluated by measuring their execution time. These performance values were continuously gathered together with characteristic data about the architecture and the functions as well as the used optimizations. Finally, this data base could be exploited to build improved optimizers.

All of these works have in common that an ACET reduction was targeted. Procedure cloning was not yet discussed in the context of compilation and optimization of embedded real-time systems. After a brief introduction to this standard optimization, the following sections explore the potential of cloning regarding WCET minimization.

4.3.3 Standard Procedure Cloning

Procedure cloning belongs to the class of inter-procedural compiler transformations where the optimizing compiler generates a specialized copy of the original procedure. Afterwards, the original function calls are replaced by calls to the newly created clones. Since constant argument values are propagated into the function body of the clones, the calling overhead for parameter handling is decreased. Moreover, the transformed code provides a more beneficial basis for aggressive inter-procedural data-flow analyses [CHK93] and offers the opportunity for improved optimizations. In particular, constant propagation and folding, copy propagation and strength reduction benefit from the modified code. Also, entire paths might be eliminated when cloning yields conditions that can be evaluated by the compiler as always *false* and thus be never executed.

Figure 4.2 demonstrates transformations performed by procedure cloning. As can be seen, the function parameters n and p of the original code on the left-hand side of the figure are substituted by the constants 5 and 2, respectively, within the cloned function f1. Moreover, the function call is adjusted.

The cloned function body could be further improved by standard compiler optimizations. First, applying *strength reduction* [BGS94] allows to replace the expensive multiplication by a shift operation. Second, the propagated constant value of n

```
int f( float *x, int n, int p ) {          int f1( float *x ) {
  for ( i=0; i<n; ++i ) {                    for ( i=0; i<5; ++i ) {
    x[i] = p * x[i];                           x[i] = 2 * x[i];
    if ( i==10 ) {...}                         if ( i==10 ) {...}
  }                                          }
  return x[n];                               return x[5];
}                               ──→        }

int main( void ) {                         int main( void ) {
  //multiple calls of f( x,5,2 );            //multiple calls of f1( x );
  return f( a, 5, 2 );                       return f1( a );
}                                          }
```

Fig. 4.2 Example for procedure cloning

could be exploited to simplify the control flow graph. By exposing the value range
of the loop induction variable i, the condition if (i==10) could be evaluated at
compile time as always *false* and be removed by dead code elimination. This transformation has a positive effect on the processor pipeline since the number of control
hazards entailed with branches is reduced. Finally, the calling overhead is reduced.
The decreased number of arguments minimizes the number of required instructions
for both the caller and the callee.

Besides the improvements concerning the program run time, the optimization has
one drawback. Each specialized copy of the function body increases code size. In
general, it is also not always permitted to remove the original function even if it is
not called anymore in the optimized program. This is due to the uncertainty if the
function might not be invoked from another compilation unit not considered in the
current optimization course. If so, its removal would be illegal.

Hence, this standard compiler optimization should be used with caution, and a
trade-off between the resulting speed-up and the increased code size, especially in
the domain of embedded systems with restricted memories, should be considered.

4.3.3.1 Selection of Functions to Be Cloned

There are different strategies to define how extensively procedure cloning should
be performed. Two factors are relevant for the optimization. First, the maximum
size of the function permitted to be cloned must be specified. This parameter can,
for example, be defined by the number of source code expressions found within the
function. All functions that exceed this parameter are omitted and not considered for
procedure cloning since they may possibly result in a too large code size increase.

The second factor guides the choice of functions to be cloned by setting constraints on the occurrence of the constant arguments. It defines how frequently a
particular constant argument must occur within all calls of the function to be cloned.
For example, the user might specify that constant argument values must be present
in more than half of all function calls. If this frequency is not reached, it will not be
considered for optimization and the function will not be cloned for this parameter.

If the code size increase is crucial, the number of additionally generated functions must be kept minimal. The only candidates for cloning are functions that are called most of the time with the same constant argument. The extreme case is the choice of functions that are always invoked with the same constant value for a particular function parameter.

However, this strict policy of trying to keep the code size small would strongly restrict the use of procedure cloning for most embedded applications since the constant values for a particular function parameter rarely remain the same, but vary between a small set of constant values. The second parameter mentioned above, defining the frequency of the occurrence of identical constant values, must thus be chosen adequately to allow the application of procedure cloning at all.

Procedure cloning is performed in three stages where each function is analyzed separately. In the first step, constant arguments and the number of their occurrences for each function parameter are collected. Hereafter, the collected arguments that do not meet the specified frequency are removed and omitted for procedure cloning. This is done by counting all function calls, where the considered argument is used, and by comparing it to the number of parameter occurrences from the previous step. In the final stage, all constant arguments that were not removed are used for procedure cloning. The original function is cloned and assigned a unique function name. The specialized argument is removed from the parameter list and directly propagated into the code by replacing the parameter variables by the constant value. Finally, the original function calls within the source code are redirected to the cloned functions.

4.3.4 Impact of Standard Cloning on WCET

Primarily, the objective of procedure cloning is to reduce the ACET. Obviously, the optimization also reduces the actual WCET of the program since the code structure is improved. However, these improvements resulting from a better code quality yield a minimal reduction of both the ACET and WCET for typical embedded systems applications as will be indicated in the following section.

More important for a WCET estimation are the transformations performed by cloning which generate code that is more accessible for high-precision WCET analyses. The augmented predictability is achieved by tackling two major problems: the explicit specification of loop bounds and the elimination of infeasible paths. Both contributions of procedure cloning enhance the tightness of the WCET estimates since they result in a more accurate description of the program behavior.

Several embedded applications spend large portions of their execution time in loops. As pointed out in Example 4.1 for the *gsm* benchmark, many loops are located in functions and their number of iterations is often specified by function parameters. If these functions are invoked with varying constants that correspond to the respective parameters, the loops exhibit a strongly deviating behavior.

To statically analyze these loops, the timing analyzer must be provided manually with loop iteration counts. To preserve WCET safeness, the loops are annotated with the maximal number of iteration counts the loop is ever executed with, i.e., the annotations must represent the global maximum of iterations for this loop considered over all execution contexts. The minimal number of iterations specified in the flow fact annotations serves exclusively for consistency checks of the ILP model during path analysis and has no effect on the computation of upper execution time bounds.

Example 4.2 For the example from Fig. 4.2, the loop bound flow facts for the original code (function **f**) and the code transformed by procedure cloning (function **f1**) have to be specified as follows using WCC's EBNF grammar (cf. Sect. 3.5):

```
int f(float *x, int n, int p) {
    _Pragma("loopbound min L max U")
    for (i = 0; i < n; ++i) {
```

$$\downarrow$$

```
int f1(float *x) {
    _Pragma("loopbound min 5 max 5")
    for (i = 0; i < 5; ++i)
```

In the original code, $\mathcal{L}$ and $\mathcal{U}$ represent global minimum and maximum of loop iterations, respectively. The larger the variations between possible loop iteration counts u, with $\mathcal{L} \leq u \leq \mathcal{U}$, the higher the overestimation of the estimated WCET. In the cloned version, precise loop bounds of 5 iterations can be specified.

Yet another code simplification has a positive effect on the tightness of the WCET estimates. Loops often consist of multiple paths resulting from conditions in the code. Some paths may have the longest execution time but are not feasible according to the program semantics. If particular conditions can not be evaluated statically, a conservative timing analysis with a restricted number of distinguished contexts has to assume the worst-case scenario where each loop iteration traverses the longest path. Using procedure cloning, these infeasible paths can be eliminated in the specialized functions.

For example, the *then*-part of the if-condition of the specialized function f1 on the right-hand side of Fig. 4.2 is never executed for parameter n = 5. Advanced data-flow based techniques are capable of detecting and eliminating conditions evaluated as always *false*. Hence, infeasible paths are not considered during WCET analysis leading to tighter WCET estimations.

4.3.5 Experimental Results for Standard Procedure Cloning

In this section, the effectiveness of standard procedure cloning is demonstrated on three complex real-life benchmarks. The evaluated benchmarks stem from the MediaBench Suite [LPMS97] which comprises different applications typically found in

Table 4.1 Characteristics of evaluated benchmarks

Benchmark	Size [kByte]	# Functions	# Loops
epic	12.5	6	41
gsm	22.8	36	48
mpeg2	30.4	14	33

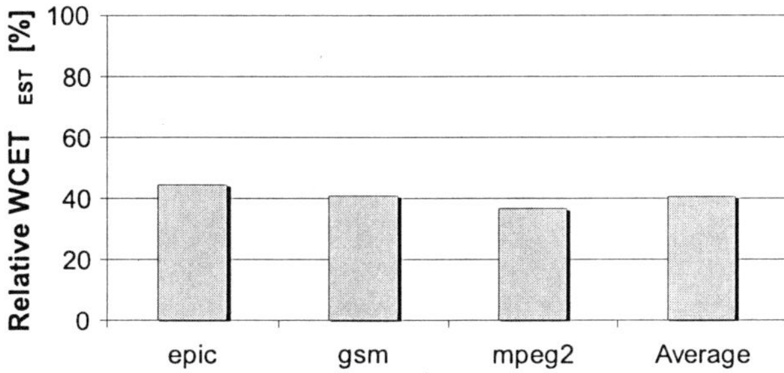

Fig. 4.3 Relative WCET estimates for standard procedure cloning

the embedded systems domain. The first benchmark is *epic*, an experimental lossy image compression utility. The already mentioned *gsm* benchmark represents an encoder and decoder for speech compression, while *mpeg2* is a motion estimation for frame pictures. Table 4.1 lists the benchmarks' size, and their number of functions and loops.

The impact of procedure cloning was studied for the TriCore processor using the WCC infrastructure. Procedure cloning was applied in the compiler frontend using the ICD-C IR optimizer. The following parameters were used for procedure cloning (see previous section): maximal function size of 2,000 expressions and a frequency of 50% (constant argument to be cloned must occur in at least half of all function calls). Due to the complexity of the timing analysis (cf. Fig. 4.1), the number of contexts was restricted to one distinguished context (no remarkable improvements were observed for the distinction of two or three contexts). Considering the run times of the WCET analysis for the original and cloned version, no significant changes could be observed.

4.3.5.1 WCET

Figure 4.3 depicts the influence of procedure cloning on the estimated WCET of the considered benchmarks. The 100% mark corresponds to the WCET estimation of the original code compiled with the standard optimizations constant folding, constant propagation and dead code elimination [Muc97]. These optimizations are applied in order to simplify the code structure and to remove dead code which can be detected even without cloning. The same optimizations are applied before and after cloning. That way, the explicit impact of cloning can be explored. On average, the estimated WCET could be reduced by 59.4%.

The estimated WCET for *epic* could be decreased by 55.6%. This is due to the code structure of *epic*: it contains a filter function consisting of 32 nested, partially data-dependent loops that is highly appropriate for cloning. Due to this code structure, cloning has a dramatic impact on WCET estimates since exact flow facts for the loops of this filter function representing the hot spot of *epic* are provided for static WCET analysis.

The *gsm* benchmark contains the already presented residual signal filter. This function is called with strongly varying constants (between 13 and 120) defining the number of iterations for its loop. Cloning creates specialized function versions for the frequently used constants. Due to the improved code analyzability after procedure cloning, the loops can be annotated with exact flow facts. This has a positive effect on the estimated WCETs. A reduction of 59.2% compared to the WCET of the non-optimized code was achieved.

The benchmark *mpeg2* contains two functions that were optimized by procedure cloning. The first function implements the full-search motion detection algorithm. It is invoked with two different constant values (8 and 16) defining the height of the frame block. Within this function, another procedure is called computing the distance between these blocks. It is invoked with the same block height constants as passed to its caller. These values are used to control the number of iteration counts for multiple loops. After cloning, the code contains a dedicated version of the full-search implementation for each block size. The loop bounds in the nested functions can again be defined more precisely, which is automatically carried out by WCC. The WCET estimate after procedure cloning is reduced by 63.4% compared to the non-optimized code.

4.3.5.2 ACET

To examine the impact of procedure cloning on the ACET, code before and after procedure cloning was measured using a cycle-true instruction set simulator. The improvements of cloning were negligible. On average for all benchmarks, an ACET decrease of 2.1% was achieved. The small degree of ACET reduction is due to the fact that cloning did not enable sufficient optimization potential for other optimizations. The comparison between ACET and WCET results highlights the significantly different impact of this compiler optimization on the performance metrics ACET and WCET.

4.3.5.3 Code Size

Finally, the code size is examined. As stated previously, the benefits of procedure cloning come at the cost of a code size increase. In particular, large function bodies or functions that are cloned multiple times for different constant arguments may exhibit a high code expansion.

Figure 4.4 depicts the relative code size, with 100% corresponding to the size of the benchmarks before cloning. The code size of the *epic* benchmark rose to 793.7%

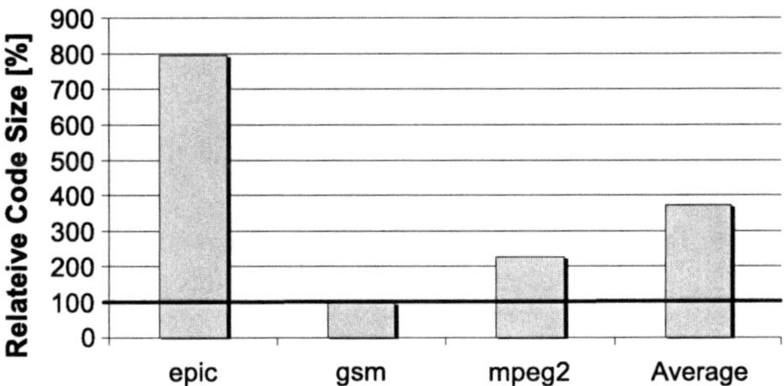

Fig. 4.4 Relative code size for standard procedure cloning

since a large function with 32 nested loops was cloned multiple times. The size of *gsm* was almost unchanged since only few small functions were cloned. Finally, a code size increase of 127.0% was observed for *mpeg2* compared to the code without extensive cloning. The average code size increase amounts to 273.2%.

For embedded systems with restricted memory resources, high code size increases, such as observed for *epic*, might not be acceptable. Therefore, a trade-off between predictability and code growth has to be taken into account and cloning of inappropriate functions should be possibly omitted.

4.3.6 WCET-Aware Procedure Cloning

Previous sections have highlighted the benefits of procedure cloning on the predictability of the code but also indicated that the standard optimization may not be suitable for embedded systems due the emerging code expansion.

The general problem of standard cloning for an improved WCET estimation is the selection of cloning candidates. Without the notion of timing in the compiler, the optimizer transforms each function that satisfies a given prerequisite, i.e., calling the function with frequently occurring constant arguments. However, cloning of functions without data-dependent loops is not promising since marginal WCET improvements with a simultaneous extensive code increase can be expected.

To cope with this problem, procedure cloning is extended by heuristics that respect the trade-off between predictability and code size. The basic idea behind the novel WCET-aware procedure cloning developed in this book is to evaluate the impact of a function transformation w.r.t. its benefits on the WCET estimation and the resulting code growth. To enhance predictability effectively, only functions on the WCEP are involved. Moreover, cloning begins with the optimization of the function that promises the highest WCET improvement. In order to control code expansion, the user may specify the maximally permitted code size increase. Therefore, taking this parameter into account allows an efficient exploitation of the available free memory for a more predictable program code.

The workflow of WCET-aware procedure cloning is outlined in Algorithm 4.1. The optimization is provided with the program $\mathcal{P}$ to be transformed and a user-defined parameter *maxFactor* representing the maximally permitted code size in-

---

Algorithm 4.1 Algorithm for WCET-aware procedure cloning

Input: $\mathcal{P}, maxFactor$
Output: $optimized\ \mathcal{P}$
  1: $maxSize \leftarrow codeSize(\mathcal{P}) \cdot maxFactor$
  2: **repeat**
  3:    **for all** $F \in WCEP(\mathcal{P})$ **do**
  4:      **if** F is cloning candidate **then**
  5:        $\hat{\mathcal{P}} \leftarrow \mathcal{P}$
  6:        $\hat{F} \leftarrow \hat{\mathcal{P}}(F)$
  7:        $cloning(\hat{F})$
  8:        $updateFlowFacts(\hat{F}) \wedge removeInfeasiblePaths(\hat{F})$
  9:        $WCETAnalysis(\hat{\mathcal{P}})$
10:        $codeSize(\hat{\mathcal{P}})$
11:      **end if**
12:    **end for**
13:    $calculateProfits()$
14:    $F_{fittest} \leftarrow fittest(\mathcal{P}, maxSize)$
15:    **if** $F_{fittest} \neq \emptyset$ **then**
16:      $cloning(F_{fittest})$
17:    **end if**
18: **until** $F_{fittest} == \emptyset$
19: **return** $\mathcal{P}$

---

crease. The algorithm consists of three phases. In the first phase (lines 3–12), a virtual cloning of copied functions $\hat{F}$ (lines 5–6) that lie on the WCEP is carried out. To cope with possible WCEP switches, cloning is repeated as long as functions on the WCEP are found. This way, even functions that become part of the longest path after the successive cloning are covered.

In the second phase (line 13), the impact of cloning on functions on the WCEP is evaluated. The parameters of function F must satisfy at least one of the following conditions such that F is considered for WCET minimization:

- used inside a loop exit condition to allow a definition of more precise loop bound flow facts,
- used inside a conditional expression so that possibly infeasible paths can be removed,
- passed as argument to a callee of F. Cloning for this case does not directly influence the WCET estimation but provides constant arguments for the callees that in turn might profit from later cloning.

The evaluation of F is performed by virtually cloning the copy $\hat{F}$ of F. Within the copy, flow facts of the data-dependent loops are automatically adjusted using a static loop analysis and WCC's flow fact manager (cf. Fig. 3.1 on p. 27). In addition, infeasible paths are eliminated. The WCC framework uses polytope models for the detection of conditions that are evaluated as always *false* [FM03].

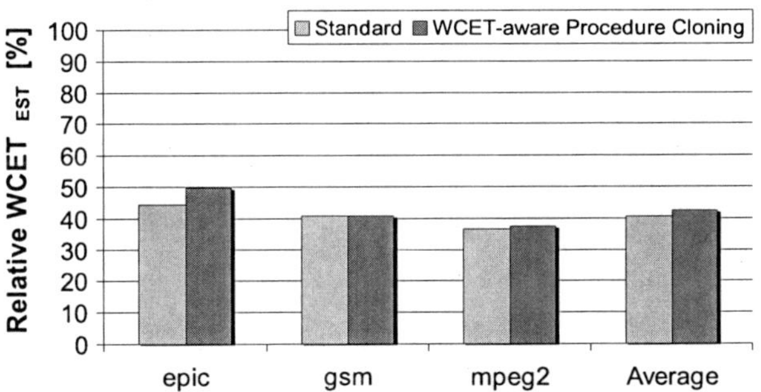

Fig. 4.5 Relative WCET estimates for standard & WCET-aware procedure cloning

Based on a comparison between the original program $\mathcal{P}$ and $\hat{\mathcal{P}}$ containing $\hat{F}$, the *cloning profit* Φ^F_{clone} can be computed:

Definition 4.2 (Cloning profit) Let $\text{WCET}_\mathcal{P}$, $\text{WCET}_{\hat{\mathcal{P}}}$ be the WCET and $\text{CS}_\mathcal{P}$, $\text{CS}_{\hat{\mathcal{P}}}$ the code size of the original and partially cloned (with function $\hat{F}$) program, respectively. The *cloning profit* Φ^F_{clone} is computed as follows:

$$\Phi^F_{clone} = \frac{\text{WCET}_\mathcal{P} - \text{WCET}_{\hat{\mathcal{P}}}}{\text{CS}_{\hat{\mathcal{P}}} - \text{CS}_\mathcal{P}}$$

Data about the WCET and the code size are made available within ICD-C via WCC's back-annotation. If there is a single function clone that replaces the original function, the size of the optimized code might get smaller resulting in a negative denominator. In that special case, the denominator is set to 1.

In the last phase (lines 14–18), cloning is performed on the fittest function $F_{fittest}$ that is assigned the highest profit $\Phi^F_{clone} > 0$. After the transformation, the optimization starts again in the first phase to capture possible WCEP switches. Algorithm 4.1 warrants the maximally permitted code size increase by considering *maxSize* and possibly excluding F from cloning if code restrictions would be exceeded.

4.3.7 Experimental Results for WCET-Aware Procedure Cloning

To show the benefits of WCC's WCET-aware procedure cloning, experiments are conducted on real-life benchmarks and compared with results of standard procedure cloning from Sect. 4.3.5.

4.3.7.1 WCET

Figure 4.5 shows the impact of standard procedure cloning (first bars per benchmark) and WCET-aware cloning (second bars per benchmark) on the estimated

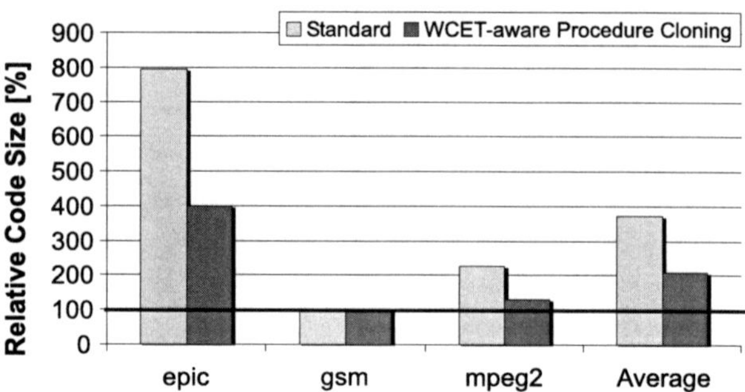

Fig. 4.6 Relative code size for standard and WCET-aware procedure cloning

WCET. The parameter *maxSize* from Algorithm 4.1 was set to 4. Again, the 100% corresponds to the WCET estimates of the original code compiled with the standard optimizations constant folding, constant propagation and dead code elimination. On average, the estimated WCET could be reduced by 57.5% for the considered benchmarks using WCET-aware cloning. As can be seen, the achieved results for both optimizations are almost identical. For all three MediaBench benchmarks, standard cloning could accomplish a slightly smaller WCET compared to the WCET-aware cloning. On average, standard cloning outperforms WCC's novel cloning merely by 1.9%.

The reason for this minimal advantage of standard cloning is that the optimization transforms even functions that do not allow a more precise specification of flow facts. Hence, standard cloning specializes more functions than the WCET-aware optimization. The resulting program code exhibits a slightly improved performance since it benefits from the traditional advantages of cloning such as a reduced calling overhead. However, as already seen for the ACET results of standard cloning, these benefits are typically negligible.

4.3.7.2 Code Size

The most interesting question is whether WCET-aware cloning can control the significant code size increase entailed with extensive specialization of functions. In Fig. 4.6, the resulting code size for standard and WCET-aware cloning is depicted. Analogous for the previous WCET results, 100% corresponds to the code size of the non-optimized code and *maxSize* is set to 4. As can be seen, the WCET-aware cloning could significantly reduce the code expansion for *epic* and *mpeg2*.

For *epic*, the code expansion of 693.7% could be cut down to 298.6%, resulting in an absolute code size of 39.42 kByte. Such a code size reduction is important for resource-restricted systems. For example, the optimized code could be allocated into TriCore's program scratchpad having a size of 48 kByte while the code after standard cloning would exceed the scratchpad capacity. Experiments were also conducted with *maxSize* = 2, that is, maximally a doubling of the code size was permitted. In this case, the relative WCET estimation of 49.6% (for *maxSize* = 4) increased to 83.4%. This result points out that some crucial functions for a more precise WCET analysis were not specialized as they would increase the code size too heavily.

Using *maxSize* = 4, similar results were observed for *mpeg2*. The relative code size could be decreased from 227.0% (for standard cloning) to 131.6% using the WCET-aware cloning. On average for all three benchmarks the relative code size reduction amounts to 209.5% (cf. bar labeled with *Average*).

These results allow to draw the conclusion that WCC's WCET-aware procedure cloning exploits the benefits of cloning for an improved WCET estimation but also keeps code expansion small by avoiding cloning of functions that do not promise a noticeably positive effect on the timing analysis. While standard cloning achieved an average WCET reduction of 59.4% at the cost of a code size increase of 273.2%, the WCET-aware optimization reduced the estimated WCET by 57.5% with a simultaneous code size increase of only 109.5%. Moreover, it could be also seen that a too heavy restriction of code expansion may prevent a precise WCET estimation.

4.3.7.3 Optimization Run Time

The optimization run time strongly depends on the number of functions that are potential candidates for cloning since they are all evaluated w.r.t. their influence on the WCET, making multiple runs of the WCET analyzer necessary. For the *gsm* benchmarks with few functions, the optimization time of WCET-aware cloning amounts to 37 minutes on an Intel Xeon 2.4 GHz system with 8 GB RAM. Most time was spent on the optimization of *epic* with 221 minutes caused by the large number of evaluated functions. However, the structure of this application is exceptional and does not represent the general program structure found in embedded system applications. On the other hand, WCET optimizations are not performed as frequently as standard optimizations on general-purpose systems but are run once to generate the final production code. Thus, the optimization run time is not a key issue and longer analysis times are acceptable.

The techniques presented in this section led to the publications [LFSP07, LFS+07, LFMT08].

4.4 Superblock Optimizations

The previous sections have demonstrated the effectiveness of compiler-based WCET minimization. However, the full optimization potential can often not be explored by numerous compiler optimizations since they are considerably limited by basic block boundaries found in the application code. To overcome this problem, a program structure called *superblock* has been introduced. It comprises several basic blocks and allows optimizations across block boundaries. This technique was thoroughly studied in the past for ACET minimization and substantial program speedups were reported [CMH91, CMW+92, CL00, KH06]. To find promising block candidates for superblock formation, block execution counts are required. For ACET minimization, profiling typically identifies assembly blocks on the most frequently executed path within the program's control flow graph.

In this book, the concept of superblocks is exploited for the optimization of embedded real-time systems that have to meet stringent timing constraints. Unlike profiling-based ACET optimization, WCC's superblock formation is driven by WCET data. Moreover, superblocks, which were constructed in the past exclusively at assembly level, are translated for the first time to source code level. Such an approach enables an early code restructuring in the optimization sequence providing more optimization opportunities for subsequent transformations. To benefit from the new constructs, the widely used compiler optimizations common subexpression elimination (CSE) and dead code elimination (DCE) are re-designed to operate on WCC's WCET-aware superblocks. This adaption allows an effective WCET minimization as code optimization is focussed on the most promising portions of the WCEP.

This section begins with a motivating example to illustrate the benefits of superblocks for compiler optimizations. Section 4.4.2 provides a survey of related work. In Sect. 4.4.3, the traditional concept of superblocks at assembly level is presented and extended towards WCET-aware source code superblocks. Section 4.4.4 describes how these superblocks are combined with the compiler optimizations CSE and DCE. Finally, experimental results obtained using these optimizations are discussed in Sect. 4.4.4.

4.4.1 Motivating Example

Traditionally, local optimizations can often not be applied across basic block boundaries, thus their optimization scope is significantly limited. Extending the optimization scope to functions or even the entire program may help to overcome these limits. However, intra- and inter-procedural optimizations must consider each CFG path equally in order to preserve program semantics. As a consequence, crucial paths through the CFG can not be aggressively optimized since optimization opportunities are inhibited by less important paths that have a minimal impact on the program performance. An effective optimization is infeasible unless the *distracting* paths are systematically excluded from the analysis.

As an example, consider the code and its CFG shown in Fig. 4.7. The expensive division `arg / 3` is computed twice on the WCEP. Thus, it is a candidate for the compiler optimization *common subexpression elimination (CSE)* that replaces multiple identical occurrences of an expression by a single variable that holds that computed value [Muc97]. However, the application of CSE is inhibited for the division in Fig. 4.7(a) since the variable `arg` does not remain unchanged between the two evaluations of expression `arg / 3` if the *else*-part of the `if`-condition is taken. Hence, the WCEP can not be optimized by CSE.

Transforming the code as shown in Fig. 4.8 enables the application of CSE. The basic block holding the statement `j = arg / 3;` is duplicated, turning

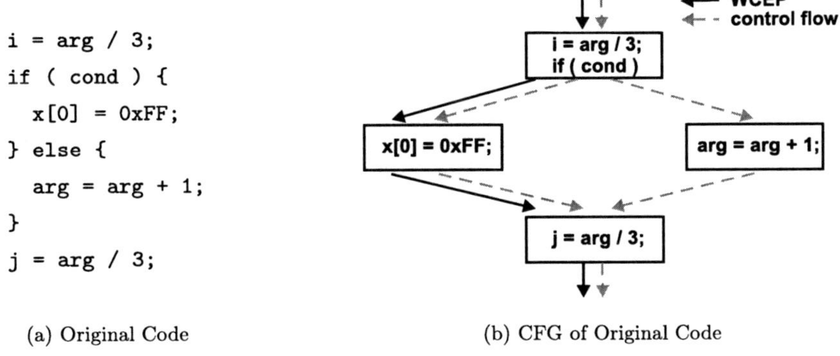

```
i = arg / 3;
if ( cond ) {
  x[0] = 0xFF;
} else {
  arg = arg + 1;
}
j = arg / 3;
```

(a) Original Code

(b) CFG of Original Code

Fig. 4.7 Example for inhibited optimization opportunity

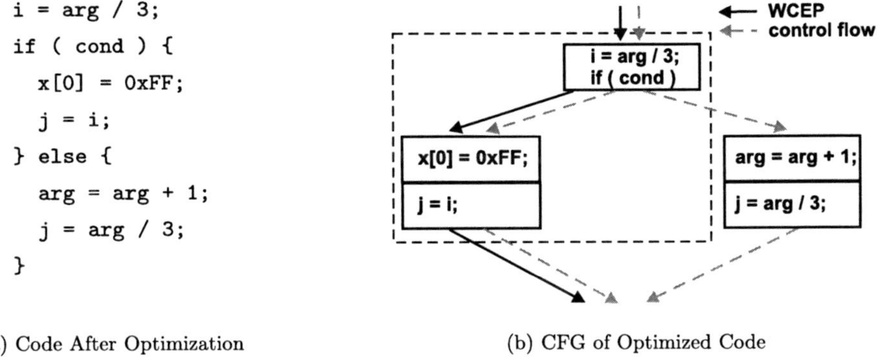

```
i = arg / 3;
if ( cond ) {
  x[0] = 0xFF;
  j = i;
} else {
  arg = arg + 1;
  j = arg / 3;
}
```

(a) Code After Optimization

(b) CFG of Optimized Code

Fig. 4.8 Example for combination of superblock formation and CSE

`arg / 3` into a common subexpression since it can not be reached from the mutually exclusive path traversing the *else*-part. Based on the restructured CFG, the expensive division can be replaced by the previously computed result stored in variable `i`. This way, the length of the WCEP is shortened at the cost of other paths leading to a decreased WCET estimation. In literature, this transformation is called *superblock formation*. In Fig. 4.8(b), the generated superblock is marked by the dotted box.

This example shows that optimizations combined with a superblock formation driven by WCEP information can find more opportunities than traditional code optimizations. In the following, this idea will be pursued in more detail.

4.4.2 Related Work

Superblock-Based ACET Minimization The concept of superblocks belongs to the class of compiler optimizations that has been extensively studied for ACET reduction. Superblocks are based on the model of *traces* which represent the most frequently executed paths in the program. The initial idea was introduced by Fisher [Fis81] who considered traces as extended regions in the code to perform instruction scheduling across basic blocks. An improved version of trace scheduling w.r.t. the compilation time and code size increase was presented in [SDJ84]. Different trace selection algorithms were evaluated in [CH88] and one of the first successful integrations of a trace scheduler into the commercial compiler Multiflow was presented in [LFK+93].

The main drawback of trace scheduling is the arising overhead for the insertion of *compensation code* after scheduling a trace to preserve program semantics. To overcome this complex *bookkeeping*, Chang introduced superblocks [CMH91] which allow an easier instruction scheduling. Moreover, this work discusses ideas for the exploitation of superblocks by standard optimizations. These ideas serve as motivation for the work presented in this section. Cohn [CL00] presented an extension to Chang's approaches w.r.t. superblock-based loop optimizations, while software pipelining operating on superblock loops was discussed in [CMW+92]. To increase the size of superblocks, Hwu proposed enlarging optimizations [HMC+93]. All the presented works have in common that the optimizations are applied in the compiler back-end at assembly level and rely on average-case execution counts of basic blocks.

WCET Minimization   The only work considering superblocks for WCET minimization was published by Zhao [ZKW+05]. This paper is most related to this work but also significantly differs in several ways. Most importantly, Zhao performs the superblock formation at assembly level while WCC's superblock construction is performed early in the compiler's optimization process at source code level. Thus, WCC's approach enables further potential for subsequent optimizations as shown in [KH06]. Moreover, in [ZKW+05] no novel superblock-based WCET-aware optimizations were developed while this book proposes a novel combination of WCET-aware superblocks with the optimizations CSE and DCE. In addition, Zhao used small programs for the evaluation of his approach, thus it is not clear if his approach also scales well for realistic applications. In contrast, the evaluation of WCC's superblock optimizations is conducted on large real-life benchmarks, which represent applications used in industry. As will be shown later, the optimizations can be carried out in an acceptable amount of time. Finally, Zhao tackles the problem of a switching WCEP in a simplified way by performing a WCET analysis after each code transformation. This approach is time-consuming and not suitable for larger applications. In contrast, WCC's superblock optimizations rely on a fast ILP-based recomputation of the WCEP after a code modification.

4.4.3 WCET-Aware Source Code Superblock Formation

In this section, the required steps for the WCET-aware formation of superblocks are discussed. Section 4.4.3.1 introduces traces as the underlying structure of superblocks and presents an improved trace selection algorithm compared to existing approaches. Based on the previous section, the concept of superblocks as well as the differences between assembly and source code superblocks are described in Sect. 4.4.3.2. Finally, an algorithm for the formation of WCET-aware source code superblocks is provided in Sect. 4.4.3.3.

4.4.3.1 Trace Selection

Traces are the underlying structure of superblocks.

Definition 4.3 (Trace) Given a control flow graph $G = (V, E)$ with nodes V corresponding to basic blocks and edges E connecting two nodes $v_i, v_j \in V$. A *trace* T is a sequence of basic blocks $T = (b_a, \ldots, b_k)$, such that for $a \leq i < k$, $(b_i, b_{i+1}) \in E$. If there is a loop L, with $\exists v_k \in L : v_k \in T$, then T is restricted by the respective loop boundaries, i.e., T does not span across basic blocks that lie outside L.

The problem of trace selection is formally defined as follows:

Problem 4.1 (Trace selection) Given a weighted control flow graph G and a set $\mathcal{T}$ of already selected traces, the problem is to find a new trace T_n such that $\nexists T_k \in \mathcal{T} :$ $T_k \cap T_n \neq \emptyset$ holds and the sum of weights of all nodes or edges within trace T_n is maximized.

In literature, the trace selection problem is handled by greedy heuristics. Popular trace selection algorithms [CH88] rely on execution counts of basic blocks or control flow edges between blocks which are typically expressed by block weights $w(b_i)$ or edge weights $w(e_i)$, respectively. For a trace T_i, both approaches begin at a block b_{start} having the highest execution count $w(b_{start})$ and being not part of any other trace T_j. In the following steps, the trace $T_i = (b_a, \ldots, b_{start}, \ldots, b_k)$ is alternately extended at both ends. With *Traces* denoting the set of already selected traces and $\delta^+(b)$ denoting the set of outgoing edges from block b, the two trace selection algorithms for the expansion of a trace at its end (extension at the trace beginning works equivalently) operate as follows:

- **Selection via node weights**: Select b_{new} such that edge $(b_k, b_{new}) \in E$, $\forall T_j \in$ *Traces* $: b_{new} \notin T_j$ and $w(b_{new}) = \max\{w(b_i) \mid (b_k, b_i) \in E\}$.
- **Selection via edge weights**: Select b_{new} such that edge $e_{new} = (b_k, b_{new}) \in E$, $\forall T_j \in$ *Traces* $: b_{new} \notin T_j$ and $w(e_{new}) = \max\{w(e) \mid e \in \delta^+(b_k)\}$, with $\delta^+(b_k) = \{(b_i, b_j) \in E \mid b_i = b_k\}$.

The trace selection based on node weights may find less suitable traces than the edge weight-based selection. This is illustrated in Fig. 4.9 which shows a fragment of a CFG with blocks and edges labeled with their execution counts which can be delivered by profiling or a WCET analysis. Starting at the if-condition, the selection based on node weights would select *blockA* for trace expansion due to the

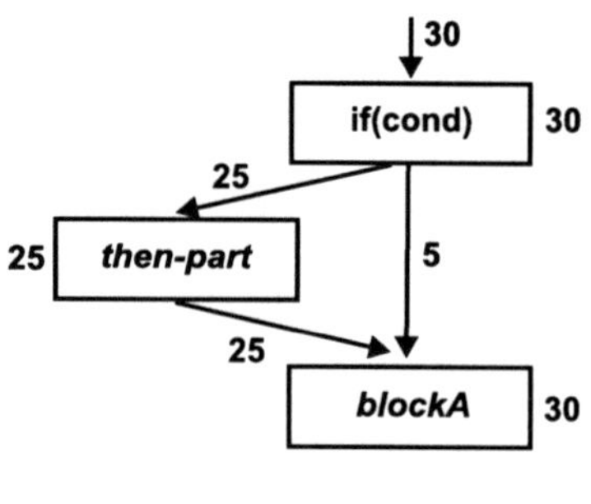

Fig. 4.9 Drawback of node weight-based approach

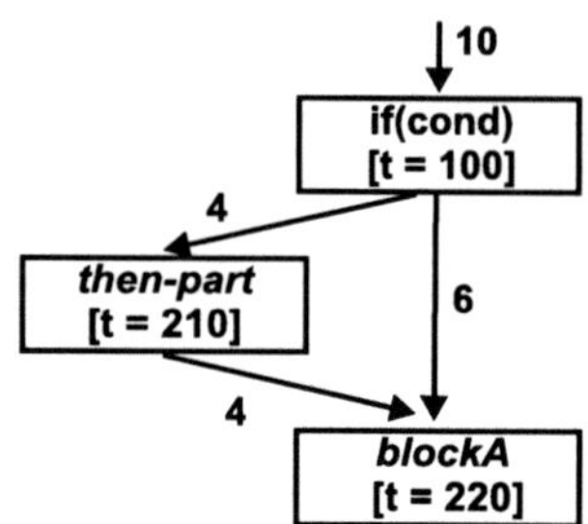

Fig. 4.10 Failure of existing trace selection approaches

higher execution counts of 30. However, as the edge weights indicate, this decision is unsuitable since the control flow traversing the *true*-edge to the *then*-part is executed more frequently than the *false*-edge. Thus, selecting the *then*-part as done by the edge weight-based approach is likely to provide a higher optimization potential.

Longest Path Approach   But even the edge weight-based selection algorithm may take inappropriate decisions leading to a trace that does not enable full optimization opportunities. The general problem with the two presented greedy algorithms is their limited, local view on the program's CFG which may miss the construction of good traces and hence promising superblocks.

Consider the weighted CFG in Fig. 4.10 where blocks are annotated with their (worst-case) execution times t (in parentheses) and edges with the execution counts. Starting at the `if`-condition, both the node and the edge weight-based approach would select *blockA* for trace expansion. Such a trace following the *false*-edge comprises code that *consumes* $6 * 100 \ cycles + 6 * 220 \ cycles = 1,920 \ cycles$. However, if the *true*-edge were taken, the trace would comprise all three blocks with a length of $4 * 100 \ cycles + 4 * 210 \ cycles + 4 * 220 \ cycles = 2,120 \ cycles$. Obviously, focusing on the longer trace promises more optimization potential.

The selection of a trace based on the *longest path* outperforms the greedy algorithms. However, it requires precise information about the execution time of each basic block. Typically, profilers used for ACET minimization do not provide this data, thus the computation of the longest path is not feasible in practice. In contrast, WCET-aware compilation, which relies on a tight integration of a static WCET analyzer into a compiler framework, has access to a precise worst-case timing data including the block WCETs.

In the following, the novel algorithm for trace selection based on the longest path is presented in an informal manner. The algorithm requires WCET estimations for each basic block at source code level. This data is provided by WCC's back-annotation (cf. Sect. 3.7).

If the starting block is located within a loop, the algorithm finds a trace T within this loop. Otherwise, an entire function is used for trace selection. The following algorithm describes the former case—the function-wise selection works analogously. The following steps are required to find the longest path:

1. Find block b_{start} with the maximal WCET serving as a heuristically promising starting point for the trace.
2. For loop L, with $b_{start} \in L$, a directed, acyclic graph $G_L = (V_L, E_L)$ is constructed, such that all blocks in L that have the same loop depth as L are added to V_L. Inner loops I of L are represented by a special node b_{loop}^I and are added to a set V_{loop}. The set of edges E_L contains all edges between blocks $b \in V_L$. Moreover, each edge (v_i, v_j), with $v_i \notin I$ and $v_j \in I$ is replaced by an edge (v_i, b_{loop}^I) and is added to E_L. The symmetric case is handled analogously. In contrast, edges inside inner loops V_{loop} as well as edges to blocks outside L do not belong to E_L.
3. L is entered through the source block denoted as b_{source}. Each block having successors outside L is called b_{sink}^i. Furthermore, a distinguished node $b_{supersink}$ is created and connected with all b_{sink}^i.
4. To find the longest path in G_L, the WCEP computed by a context-sensitive timing analyzer is not suitable as trace since it may contain code on multiple mutually exclusive paths executed in different contexts. To this end, the widely used *implicit path enumeration technique* (IPET) (cf. Sect. 2.3.4 on p. 19) is applied to find the longest path in G_L where for each branch only one path is selected. In this approach, an integer linear program is formulated by translating the control flow in G_L into a system of linear constraints. The objective function represents the execution time of possible paths in G_L. Maximizing this function under the given constraints yields the longest path in G_L. Due to the small number of possible blocks to be considered, the run time of the IPET approach is typically negligible amounting to few seconds.
5. The final trace is selected by starting at b_{start} and adding alternately one block from the longest path to the beginning and end of the trace. For each lengthening of the trace, WCC estimates the code size expansion that would occur if this trace was used for superblock construction. The estimated code size is compared with a user-defined code expansion restriction to avoid extensive code increases.
6. This algorithm is repeated as long as there are unprocessed blocks which are not part of any trace and have a WCET > 0. Thus, the algorithm also handles blocks in inner loops of L and blocks outside L at some point.

4.4.3.2 Concepts of Superblocks

The motivation for the development of superblocks originates from the idea to simplify instruction scheduling performed on traces. Trace scheduling was initially developed by Fisher [Fis81] to enable a more efficient scheduling across basic block boundaries. Code motion below *side exits* (conditional branches out of the middle of the trace) can be handled in a straightforward manner by copying the moved instructions into the side exits. Handling of instructions moved above side exits (speculative execution) is also simple. Complex *bookkeeping* is required when code is moved below or above *side entrances*, which are defined as a block $b_{in} \in T$ with a predecessor $b_p \in T$ and at least one predecessor $b_{off} \notin T$.

The bookkeeping associated with side entrances can be avoided if they are removed from the trace. For this purpose, superblocks were developed at assembly level.

Assembly Superblocks   The concept of superblocks was published for the first time by Chang et al. [CMH91]. The authors describe the structure of a superblock as a trace that has no side entrances:

Definition 4.4 (Superblock) A *superblock* S is a trace which can be entered only at the first basic block b_{start}, i.e., S is a trace $T = (b_{start}, \ldots, b_{end})$ such that $\forall b \in T \setminus \{b_{start}\} : \forall (b_i, b) \in E : b_i \in T$. In contrast, control from S may leave at one or more exit points.

Since the size of *natural* superblocks found in the program code is typically small, superblock enlarging optimizations [HMC+93] are used. The main technique is *tail duplication* which eliminates side entrances of arbitrary traces. For a side entrance b_{in} in a trace $T = (b_a, \ldots, b_{trc}, b_{in}, \ldots, b_e)$, a copy of the tail portion of the trace from the side entrance to the end $b_{in}, \ldots, b_e$ is created and all side entrance edges $e \in \{(b_{pred}, b_{in}) \in E \mid b_{pred} \neq b_{trc}\}$ are redirected to the corresponding duplicated basic block b'_{in}. Figure 4.11 illustrates the superblock formation at assembly level. On the left-hand side of the figure, a trace is selected (marked in bold). Using tail duplication, which copies block $L3$, a superblock is built (cf. dotted box in Fig. 4.11(b)). The superblock formation at this abstraction level of the code is easy since merely block labels have to be adjusted, while the code in the duplicated tail portions remains the same as in the original code.

Source Code Superblocks   The definition of basic blocks at source code level is equivalent to its counterpart at assembly level, making a translation of the superblock concepts possible. Following Definition 4.4, superblocks at source code level are defined as traces of a high-level control flow graph with no side entrances. However, WCC's source code superblocks differ in one point from the traditional definition. Unlike Chang's superblocks [CMH91], a WCC superblock $S = (b_0, b_1, \ldots, b_n)$ may contain *inner loops*. These loops are represented on the trace by their loop headers. Hence, a loop L is referred to as an inner loop if its loop header is a basic block b_{head} such that $b_{head} = b_k$ with $k \in 1, \ldots, n$, and block b_{k+1} is the next basic block after loop L. This idea of a special handling of inner loops is similar to Fischer's trace definition [Fis81] which treats loops as a single operation called *loop representative*.

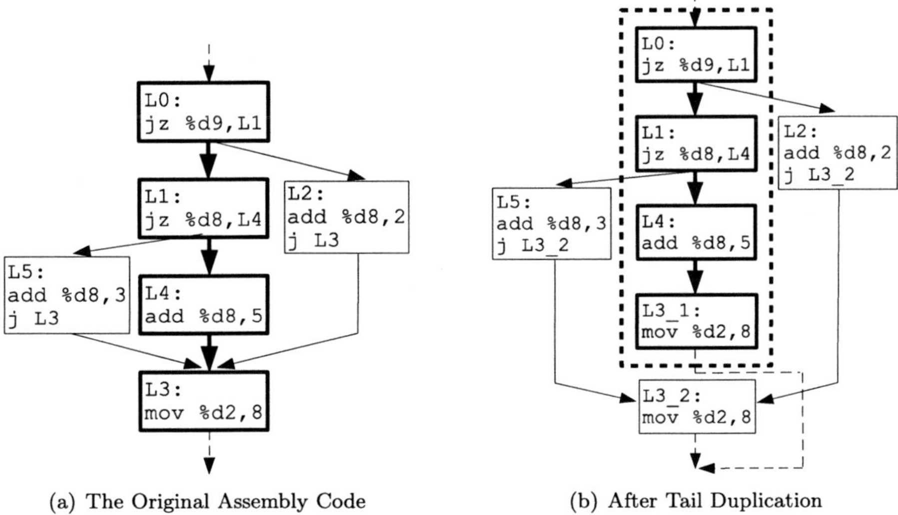

(a) The Original Assembly Code   (b) After Tail Duplication

Fig. 4.11 Superblock formation at assembly level

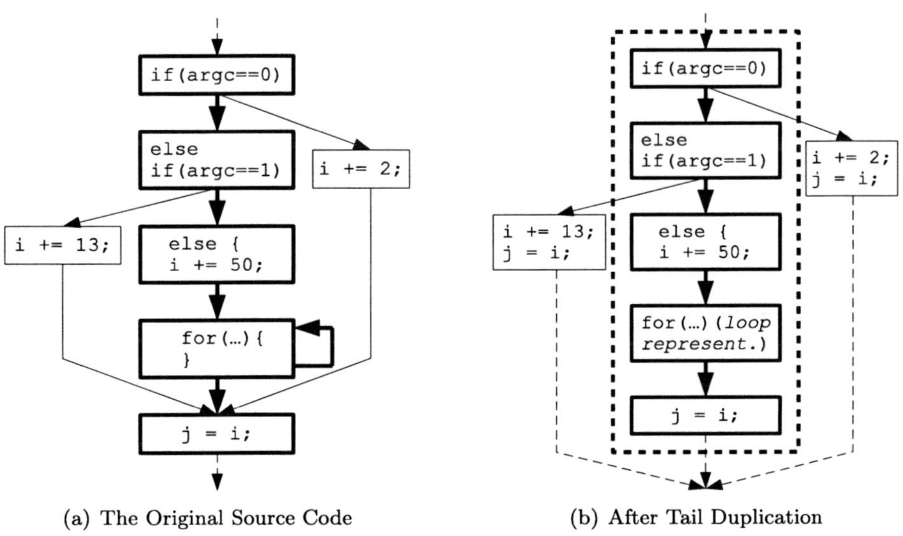

(a) The Original Source Code   (b) After Tail Duplication

Fig. 4.12 Superblock formation at source code level

An example for a superblock formation at source code level for the programming language ANSI C is depicted in Fig. 4.12. Again, the trace including an inner for-loop is marked in bold on the left-hand side of the figure. In contrast to the assembly code, basic blocks do not consist of operations but of source code statements. Moreover, basic blocks at this level are not distinguished by unique block labels, making an explicit control flow modification from one block to another arbitrary block difficult. Compared to the superblock formation at assembly level (cf. Fig. 4.11), which was easy to perform due to simple duplication of the corresponding tail portion and the adjustment of the block labels, a tail duplication at source code level is more elaborate. As can be seen in Fig. 4.12(b), the superblock formation requires a duplication and insertion of the statement `j = i` into the CFG paths other than the trace.

The triplication of statement `j = i` could be avoided if, similar to assembly level, a new basic block holding this statement would be generated and control flow using `goto`'s would be redirected to this block. However, this strategy was intentionally avoided. Many compiler optimizations rely on well-structured code and the insertion of `goto`'s would make the code unqualified for these transformations. Thus, the benefit of superblocks to enable other optimizations would be heavily decreased.

4.4.3.3 Superblock Formation

In the following, the formation of WCET-aware superblocks at source code level is discussed in more detail. The formation starts with a preprocessing step to enlarge code fragments suitable for superblocks and to remove code constructs that inhibit a superblock construction.

Preprocessing   The preprocessing phase begins with the application of the well-known optimizations *function inlining*, which replaces function calls by the corresponding callee bodies, and standard *loop unrolling* that expands loop bodies. These optimizations are also proposed for the formation of assembly superblocks in [HMC+93]. In the next step, programming language constructs that lead to *unstructured* code are tried to be eliminated since they prevent a proper superblock formation. Examples are ANSI C `goto`-statements. A detailed description for the elimination of unstructured code can be found in [EH94]. If the undesired construct can not be removed, then functions containing these constructs are omitted for superblock formation. ANSI C `switch`-statements without an explicit `default`-case also require a special handling. During the superblock formation, code following a `switch`-statement is possibly moved into the respective `cases` of a `switch`-statement. To ensure that this code is always executed (like in the original version) even when no `case` is entered, an initially empty `default`-case is added which can hold a copy of the duplicated code.

Formation Algorithm   After preprocessing, a WCET analysis of the compiled program is performed at assembly level and the WCET information is made available at source code level using WCC's back-annotation. This information allows to order functions w.r.t. their WCETs and to begin the superblock optimizations with the function exhibiting the highest WCET. Such a strategy is important for resource-restricted embedded systems since it allows to exploit maximal optimization potential before code expansion restrictions are exceeded.

Algorithm 4.2 Algorithm for WCET-aware source code superblock formation

---

Input: *WCET-aware Trace T*
Output: *Superblock S B*

```
 1: block currBB ← endNode(T)
 2: block lastBB ← endNode(T)
 3:
 4: /* Iterate trace, starting at the end. */
 5: while currBB ≠ startNode(T) do
 6:    /* Perform tail duplication. */
 7:    if |δ⁻(currBB)| > 1 then
 8:       for all predBB ∈ δ⁻(currBB) do
 9:          if predBB == tracePredBlock(currBB) then
10:             moveStmt(firstStmt(currBB), lastStmt(lastBB), lastStmt(predBB))
11:          else
12:             copyStmt(firstStmt(currBB), lastStmt(lastBB), lastStmt(predBB))
13:          end if
14:       end for
15:    end if
16:    currBB ← tracePredBlock(currBB)
17:    if isPrecededByInnerLoop(currBB) then
18:       currBB ← predBeforeLoop(currBB)
19:    end if
20:    if isConditional(lastStmt(currBB)) then
21:       lastBB ← currBB
22:    end if
23: end while
24: return  T
```

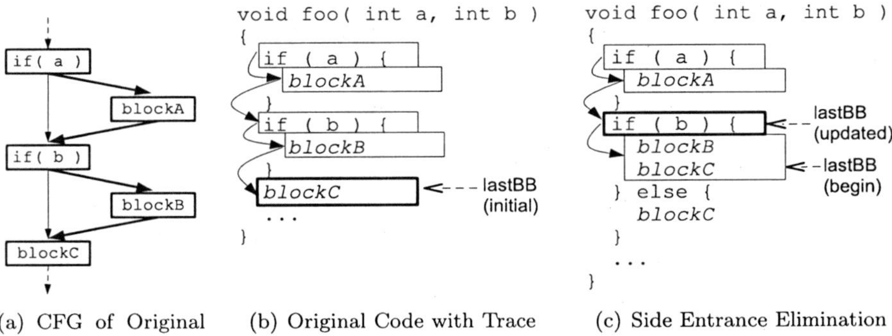

(a) CFG of Original Code with Trace   (b) Original Code with Trace   (c) Side Entrance Elimination

Fig. 4.13 Successive source code superblock formation

Source code superblock formation is depicted in Algorithm 4.2. Figure 4.13 shows an example CFG and the respective code with a trace marked by the arrows which serves for illustration of the algorithm. The basic idea of the algorithm is to traverse trace T backwards beginning at the last block and to eliminate found side entrances by duplicating the so far traversed tail portions of the trace in all CFG paths other than the trace. The original tail portion is moved behind the predecessor block on the trace to enlarge the superblock. Following this strategy, the superblock is iteratively increased by merging it with the predecessor blocks on the trace. The superblock formation continues as long as the start node of the trace T has not been reached or code size constraints have been exceeded.

Variable *currBB* represents a pointer to the basic block that is currently considered for superblock formation and used to traverse the trace backwards. In Fig. 4.13, it is marked by the bold box. To keep track of the current tail portion that must be duplicated, two pointers (*currBB* and *lastBB*, lines 1–2) are required that mark the first and last statement of the tail. If *currBB* has more than one incoming edge ($\delta^-(currBB)$), tail duplication is performed (lines 7–15). If the predecessor of *currBB* is lying on the trace, then all statements (including inner loops) between the first statements in *currBB* and last statement in *lastBB* are moved behind the last statement of the trace predecessor *predBB* (lines 9–10). Otherwise, this set of statements is copied into the other CFG paths being not part of the trace (line 12). Note that code size constraints were already checked during trace selection using the longest path approach, hence are not required to be tested again during superblock formation.

Finally, *currBB* is set to the next predecessor block on the trace (line 16), while inner loops are omitted (lines 17–19). Moreover, if the tail end *lastBB* is inside a conditional statement c, *lastBB* is moved out of c to enable a duplication of the entire conditional statement (lines 20–22).

Figure 4.13(c) shows the situation after the first side entrance elimination to *blockC*. Block *blockC* was moved onto the trace in the *then*-part of the second if-condition and copied into the non-trace predecessor (*else*-part). Moreover, *lastBB* was updated to the beginning of its embedded conditional statement.

In the next step, the superblock constructed according to Algorithm 4.2 is passed to the optimizer. After the optimization, a WCEP recomputation is required to tackle the path switch which may occur during superblock formation and its optimization.

IPET-Based WCEP Update   To make sure that subsequent optimizations do not operate on an outdated WCEP, the timing information must be updated. A frequent use of a costly WCET analysis is not feasible. Thus, WCC updates the WCEP data during superblock optimization using the IPET approach. For the underlying ILP model, which represents the program flow at source code level, the reader is referred to [Kel09]. The involved ILP model requires WCET information for basic blocks which are extracted from the initial run of the timing analyzer aiT before the superblock optimization begins. Since execution contexts are not distinguished, the solved ILP model is less complex than a full WCET analysis. Hence, the WCEP recomputation is faster but also less precise. Such trade-offs are typical for compiler optimizations where complexity is decreased at the cost of precision. It should be noted that the conducted IPET-based computation is not meant to replace a safe static WCET analysis but should be considered as a fast heuristic which helps to indicate potential WCEP switches.

For blocks that were modified by the optimizations, estimations of their WCET are made by the optimizer. The estimation uses the *tree-based approach* (cf. Sect. 2.3.4 on p. 19) which computes worst-case timing for constituents of statements based on combination rules that depend on the statement type. After some iterations (defined by the user), a full WCET analysis is performed to obtain precise timing information.

4.4.4 WCET-Aware Superblock Optimizations

In this section, the exploitation of WCET-aware superblocks for compiler optimizations is discussed. Section 4.4.4.1 briefly introduces basic techniques from static program analysis required for the development of the superblock-based optimizations common subexpression elimination and dead code elimination. The optimizations themselves are discussed in Sects. 4.4.4.2 and 4.4.4.3, respectively.

4.4.4.1 Static Program Analysis

Static program analysis tries to determine dynamic properties of the program without actually executing it. For WCC's superblock optimizations, it must be known which statements access which storage locations (variables). For high level programming languages, such as C, this analysis is not trivial when pointers are taken into account.

Alias Analysis   The goal of this analysis is to determine whether multiple memory references point to the same memory area. Based on this alias knowledge, it can be inferred which variables are affected by which statements. Compilers can use these alias relationships to simplify code. The results of an alias analysis are *points-to sets* that are attached to a program variable and indicate to which other variables it points to.

Def/Use Sets   For many compiler optimizations, it is not sufficient to know to which variables a pointer may point to, but which variables are read or written by which expressions. These results can be expressed by *def/use sets*, which represent sets of symbols from/to which an expression *may* read/write (USE_{may} and DEF_{may}). A possible computation of def/use sets, which are assigned to expressions and also integrate results of an alias analysis, is based on so-called *syntax directed definitions* [Ton99].

Livetime Analysis The last analysis required for the superblock-based optimizations is the *livetime analysis* [App97]. It is a classical data flow analysis which determines for each program point if a variable is *live* or otherwise *dead*. More accurately, a variable v is called live on a CFG edge if there is a directed path from that edge to a use of v that does not contain any redefinition of v. A variable is *live-in* at a statement s if it is live on any of the incoming edges of s. A variable v is called *live-out* at a statement s if it is live on any of the outgoing edges of s. LIVE-IN$_{\mathrm{may}}(s)$ and LIVE-OUT$_{\mathrm{may}}(s)$ are the corresponding *may*-sets at s for all v.

4.4.4.2 WCET-SB Common Subexpression Elimination

The optimization common subexpression elimination (CSE) is a well-known technique that removes recomputations of *common subexpressions* and that replaces them with uses of saved values. A common subexpression denotes an occurrence of an expression in a program if there is another occurrence of this expression whose evaluation always precedes this one in execution order and if the operands of this expression remain unchanged between the two evaluations [Muc97].

Local CSE operates on the limited scope of a single basic block. Global CSE works on entire functions but side entrances in the control flow graph often cancel opportunities for CSE since common subexpressions may be overwritten in the side entrance path. Superblock-based CSE (*SB-CSE*) can outperform the local and global CSE since it operates on multiple basic blocks and removes conflicting side entrances.

WCC's SB-CSE is similar to Ghiya's approach [GH98] and relies on def/use sets. The optimization traverses a superblock in a top-down manner and updates for each superblock statement the set of *available* expressions. Similar to the general definition of availability in the context of data flow analysis [ASU86], availability in superblocks is defined as follows:

Definition 4.5 (Available expression) In a superblock *SB*, an expression e is called *available* at statement $s \in SB$ iff e was computed in a preceding statement $s_{comp} \in SB$ and there is no re-definition of any operands $o \in e$ between the evaluation of s_{comp} and s.

To find available expressions at statement s, a list *availList* containing available expressions found during the superblock traversal is maintained. At each expression $e \in s$ it is checked if e redefines any of the operands of the available expressions *availExp* $\in$ *availList*, i.e., USE$_{\mathrm{may}}$(*availExp*) $\cap$ DEF$_{\mathrm{may}}(e) \neq \emptyset$. If so, *availExp* is removed from *availList*. Otherwise, if e equals *availExp*, a mapping *availExp* $\rightarrow e$ is stored for later replacement. If inner loops are encountered during the superblock traversal, *availList* must be possibly updated, i.e., redefined expressions within the loop are removed from *availList*. Finally, CSE traverses all collected mappings, creates a temporary variable $t = $ *availExp*, and replaces all redundant $e_1, \ldots, e_n$

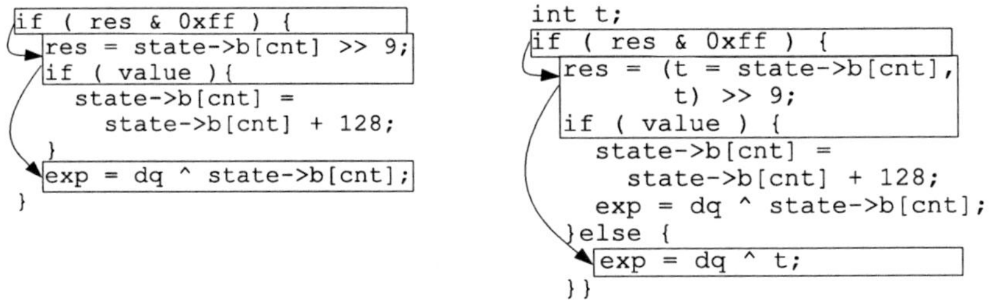

```
if ( res & 0xff ) {
    res = state->b[cnt] >> 9;
    if ( value ){
        state->b[cnt] =
            state->b[cnt] + 128;
    }
    exp = dq ^ state->b[cnt];
}
```

(a) Original Code

```
int t;
if ( res & 0xff ) {
    res = (t = state->b[cnt],
           t) >> 9;
    if ( value ) {
        state->b[cnt] =
            state->b[cnt] + 128;
        exp = dq ^ state->b[cnt];
    }else {
        exp = dq ^ t;
    }}
```

(b) After Superblock-CSE

Fig. 4.14 Example for superblock-based CSE

by t. An example from the *G.721 decoder* benchmark for SB-CSE is shown in Fig. 4.14 (trace is marked). As can be seen, the expression `state->b[cnt]` can be reused in the superblock.

4.4.4.3 WCET-SB Dead Code Elimination

Unlike CSE, dead code elimination does not just eliminate redundant computations but deletes statements completely. In case of the superblock-based DCE (*SB-DCE*), statements can be removed from a superblock by either deleting them or by moving them outside the superblock. By definition, *dead* statements s_{dead} are statements with $\text{LIVE-OUT}_{\text{may}}(s_{dead}) \cap \text{DEF}_{\text{may}}(s_{dead}) = \emptyset$. Results computed by s_{dead} are not required in the further program flow.

Definition 4.6 (Superblock-dead) A statement s_{dead}^{SB} is said to be *superblock-dead* in superblock *SB* iff it is not *dead* and none of its defined variables ($\text{DEF}_{\text{may}}(s_{dead}^{SB})$) is read in *SB* before being re-defined.

A superblock-dead statement s_{dead}^{SB} can be removed from a superblock *SB* by copying s_{dead}^{SB} into all control flow paths that exit the superblock and which contain statements that require the results of s_{dead}^{SB} [CMH91]. WCC's SB-DCE algorithm moves each superblock-dead statement s_{dead}^{SB} downwards in the superblock passing all succeeding statements. When a branch statement, i.e., a side exit, is passed, the algorithm copies s_{dead}^{SB} into all successor blocks b_{succ} which

- are not part of *SB* and
- for which s_{dead}^{SB} is live at b_{succ}, i.e., $\text{LIVE-IN}_{\text{may}}(b_{succ}) \cap \text{DEF}_{\text{may}}(s_{dead}^{SB}) \neq \emptyset$

The motion of s_{dead}^{SB} is stopped when the superblock end is reached. If s_{dead}^{SB} is live in the superblock end-block b_{end} ($\text{LIVE-OUT}_{\text{may}}(b_{end}) \cap \text{DEF}_{\text{may}}(s_{dead}^{SB}) \neq \emptyset$), s_{dead}^{SB} is kept at the superblock end to preserve data dependencies. Otherwise, s_{dead}^{SB} can be completely removed from the superblock. An example for the application

```
void foo( int arg ) {           void foo( int arg ) {
  int i = compute( arg );         int i;
  if ( cond1 ) {                  if ( cond1 ) {
    work( i );                      i = compute( arg );
  } else {                          work( i );
    work( arg );                  } else {
    if ( cond2 ) {                  work( arg );
      do( arg );                    if ( cond2 ) {
    }                                 do( arg );
} }                             } } }
```

(a) Original Code   (b) After Superblock-DCE

Fig. 4.15 Example for superblock-based DCE

of the SB-DCE is depicted in Fig. 4.15. As can be seen, the computation of `i`
could be removed from the superblock traversing the *else*-part of the outermost `if`-
condition.

4.4.5 Experimental Results for WCET-Aware Superblock Optimizations

This section evaluates the impact of WCC's WCET-aware superblock optimizations
on the WCET estimates of real-life benchmarks. In total, 55 benchmarks from the
test suites DSPstone [ZVS+94], MRTC [MWRG10], MediaBench [LPMS97], and
UTDSP [UTD10] were involved in the experiments. The experiments were conducted for the TriCore TC1796 processor with enabled I-cache.

For the experiments the following optimization parameters were used: the code
size restrictions allow the maximal superblock size to be 5 times as large as the original trace, and a maximal increase of a function and the entire program by a factor
of 3 and 2.5, respectively. Moreover, the timing analyzer aiT was run after each 4th
superblock formation including the application of the superblock optimizations to
update WCET information. For the remaining steps, the IPET-based approach (cf.
Sect. 4.4.3.3) was used. These settings were empirically determined and showed a
good performance.

4.4.5.1 WCET

Due to space constraints, Fig. 4.16 shows the impact of different common subexpression elimination strategies on the WCET estimates for a subset of 15 benchmarks. These benchmarks were chosen since they represent the typical effects of
the optimization of the considered benchmarks. The *Average*-bars on the right-hand
side of the figure indicate the results averaged for all considered 55 benchmarks.
The 100% mark, which serves as reference, represents the WCET of the benchmarks compiled with the highest optimization level (*O3*) and disabled CSE. The

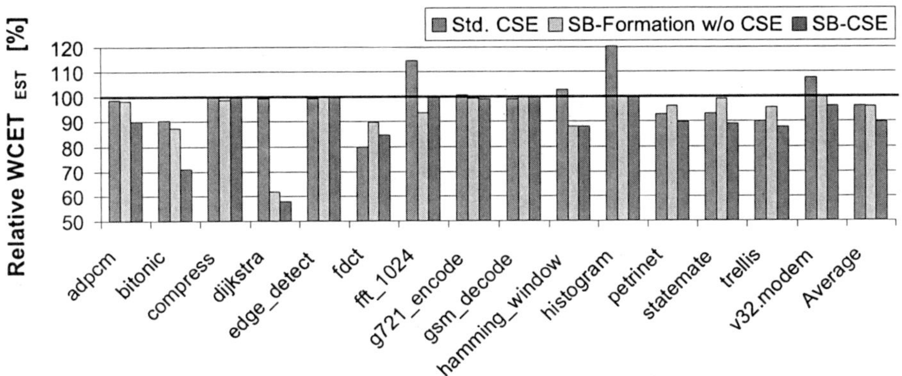

Fig. 4.16 Relative WCET estimates for SB-CSE

first bars per benchmark (labeled with *Std. CSE*) represent the results for the standard local ACET CSE. As can be seen, the WCET improvements are marginal with an average improvement of 3.4%. The second bars represent the WCET for the code optimized with *O3*, disabled CSE, and the WCET-aware superblock formation. An average WCET improvement of 4.0% was achieved. Based on these results it can be inferred that the superblock formation is often beneficial for the performance since it establishes optimization potential for subsequent standard optimizations. This observation is conform with the results in [KH06].

It should be also noted that superblock formation exploits WCC's capability to quantify transformation effects. If the estimated WCET is larger after the code modification than before, then the modification is rolled back. Thus, the diagram does not show any results larger than 100%. This capability is not available in many compilers, thus standard code transformations may yield an undesired performance degradation.

Best results were achieved for the CSE based on the WCET-aware superblocks (bars labeled with *SB-CSE*). WCET improvements of up to 42.1% for the *dijkstra* benchmark were observed. The average improvement for the 55 benchmarks amounts to 10.2%. This result is remarkable since it increases the optimization potential of a traditional, intensively studied compiler optimization by a factor of 3. From the comparison between the second and third bars also the conclusion can be drawn that superblock formation often provides optimization potential which can only be fully exploited by a tailored superblock-based optimization (cf. e.g., benchmark *statemate*). SB-CSE also exploits WCC's rollback mechanism to restore the original program if the optimization increases the estimated WCET for example due to additional spilling. Thus, WCETs larger than 100% are not observed.

The results for the superblock DCE achieved for the 15 representative benchmarks are shown in Fig. 4.17. Compared to the 100% reference mark, which represents the WCET for the highly optimized code using *O3* and disabled DCE, the standard local ACET DCE achieved on average for all 55 benchmarks a WCET reduction of 2.0%. The superblock formation amounts to an average WCET reduction

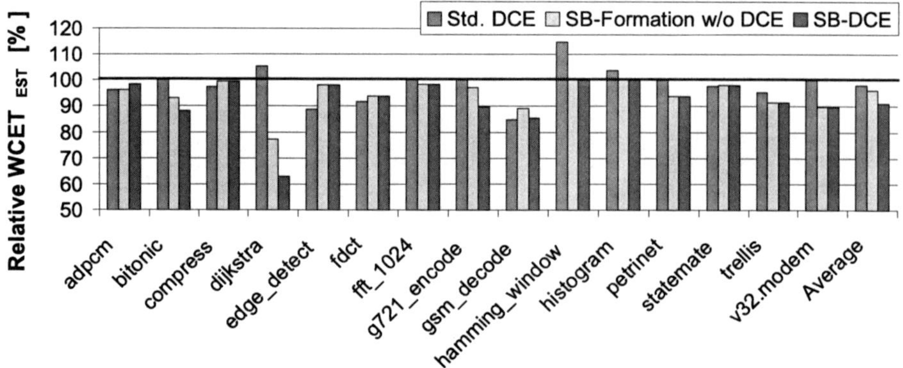

Fig. 4.17 Relative WCET estimates for SB-DCE

Table 4.2 ACET results for SB-optimizations

Optimization Level	Average ACET
O3 with Std. CSE	97.9%
O3 with SB-Formation w/o CSE	97.5%
O3 with SB-CSE	95.1%
O3 with Std. DCE	98.6%
O3 with SB-Formation w/o DCE	98.2%
O3 with SB-DCE	97.9%

of 4.0%. Again, the most effective WCET reduction was achieved with the DCE based on the WCET-aware superblocks, yielding an average improvement of 8.8%. Again, WCC's rollback mechanism was used if required.

4.4.5.2 ACET

WCC's superblock formation is driven by WCET data. Thus, it is interesting to explore its impact on the ACET. As rows one and four in Table 4.2 show, the reductions of the ACET for standard local CSE and DCE applied to the 55 benchmarks are comparable with those achieved for the WCET. This is obvious since the local CSE and DCE do not prioritize any paths, thus they have similar effects on the considered program execution times. Moreover, a comparison between the ACET and WCET results for the superblock formation and the superblock-based optimizations shows that higher improvements are achieved for the WCET reduction. One reason is that the WCET-aware optimizations focus on the WCEP which might be different from the most frequently executed path. This emphasizes the need for tailored WCET-aware optimizations which might operate differently compared to traditional ACET optimizations.

4.4.5.3 Code Size

Since the superblock formation is a code expanding transformation, the resulting code size increase is critical. The code size for SB-CSE and SB-DCE was measured for two different scenarios. In a first scenario, the same code size restrictions during trace selection were utilized as for the previously mentioned WCET and ACET experiments. For this configuration, an average code size increase of 23% for the SB-CSE and 28% for the SB-DCE was observed. Larger code size increases were typically found for smaller benchmarks with a size of few kByte, which is acceptable for modern systems.

In a second scenario, the code expansion was not limited. Code size increases for both optimizations of approximately 107% were measured on average. Simultaneously, the WCET results are slightly degraded. A possible reason might be adverse instruction cache overflows cancelling further optimization benefits. Thus, it can be inferred that a code size restriction is mandatory for a balanced trade-off between the WCET improvement and the code size increase.

4.4.5.4 Optimization Run Time

Finally, the run time of WCC's superblock optimizations on an Intel Xeon 2.4 GHz system with 8 GB RAM was measured. In a first scenario, the WCET analyzer aiT was run after each 4th superblock formation and optimization as in the previous experiments. In this configuration, the optimization run time increased by 540% compared to the time when standard optimizations are applied instead. This additional time is acceptable for code optimization of embedded systems where performance is the primary goal.

To check if no WCEP switches were missed between two runs of the timing analyzer, aiT was run after each second superblock formation in a second scenario. This led to an increase of the optimization time by 757% and marginal WCET improvements of less than 1%. Thus, it can be concluded that too frequent WCEP updates by a costly WCET analysis do not pay off.

The techniques presented in this section have been published in [LKM10].

4.5 Loop Unrolling

Program loops are notorious for their optimization potential on modern high-performance architectures. Compilers aim at their aggressive transformation to achieve large improvements of the program performance. In particular, the optimization loop unrolling has shown in the past decades to be highly effective in achieving significant increases of the ACET.

Loop unrolling is a code transformation that replicates the body of a loop a number of times and adjusts the loop-control accordingly. The number of replications is called the *unrolling factor u* and the original loop is often termed *rolled loop*. The main benefits of the optimization are a reduced loop increment-and-test overhead and an increased instruction level parallelism. The explicit parallelism is considered as the essential benefit of unrolling since it potentially enables other compiler optimizations. However, loop unrolling is also entailed with negative side-effects. The main drawbacks are an adverse impact on the instruction cache (I-cache) due to the inherent code size increase, and additional spill code due to an increased register pressure. Therefore, sophisticated control mechanisms for an effective unrolling are required which exploit the benefits of the optimization while avoiding its disadvantages.

In this book, the advanced infrastructure of the WCC framework is utilized for the development of sophisticated heuristics that focus on an explicit WCET minimization. The presented loop unrolling is completely driven by worst-case timing information which is not sensitive to input data. Hence, it is more reliable than previous ACET approaches that rely on profiling information which might change for different inputs leading to suboptimal or even adverse compiler decisions.

The remainder of this section is organized as follows. In Sect. 4.5.1, the significance of suitable control mechanisms for unrolling is motivated by a real-life example. Section 4.5.2 provides an overview of related work regarding loop unrolling. Section 4.5.3 provides a brief introduction to standard loop unrolling as found in many compilers. The novel heuristics for a WCET-aware loop unrolling are presented in Sect. 4.5.4, followed by experimental results on real-life benchmarks in Sect. 4.5.5.

4.5.1 Motivating Example

Loop unrolling belongs to the class of traditional code expanding compiler optimizations. Therefore, the compiler user is faced with the question how extensively loop body duplication should be conducted. A too careful unrolling wastes optimization potential while a too heavy transformation cancels optimization merits and even degrades program performance.

Standard compilers use simplified heuristics to decide how often a loop body is duplicated. This global decision is used for any loop in the program. However, due to different structures of program loops and their execution contexts, such as available processor registers, a global unrolling factor is not sufficient. To accomplish an effective loop unrolling, an individual assessment of each loop L is required to determine the most promising unrolling decision. This decision relies on different parameters characterizing L which are, however, usually not available in standard compilers.

The degree of unrolling at source code level can be controlled by the number of ANSI C expressions that the unrolled loop may maximally hold. This limitation serves as a common unrolling heuristic found in many standard compilers and can be set by the compiler user as a global optimization parameter.

Figure 4.18 illustrates the influence of unrolling on the WCET estimation for the *fft* benchmark from DSPstone [ZVS+94] which contains two loops being unrolling

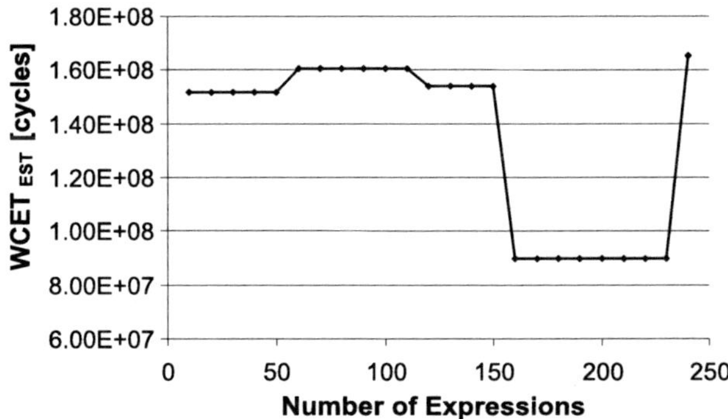

Fig. 4.18 Impact of loop unrolling on WCET estimation

candidates. The results were generated by WCC using *O3* as optimization level during code generation.

The horizontal axis represents the maximal number of permitted expressions in the body of each finally unrolled loop. Relaxing the unrolling limitation to 60 expressions allows a single duplication of the body of one of the loops, leading to a slight increase of the WCET estimation since additional spill code is introduced. Setting the limitation to 160 expressions yields a further unrolling of both loops. This transformation enables other optimizations which generate code with a decreased number of spilling instructions as a result of the reduced register pressure. This positive effect on the WCET estimation is cancelled when the loops are further unrolled at the 240-expression boundary. At this point, further I-cache misses due to a cache overflow arise.

The observations in Fig. 4.18 point out the importance of suitable unrolling heuristics. A naive unrolling without the consideration of influencing factors may even amount to a degradation of the program performance. WCC's WCET-aware loop unrolling performed at source code level exploits the benefits of loop unrolling and prevents its negative effects by integrating information from different parts of the compiler infrastructure to find a promising unrolling strategy for each individual loop.

4.5.2 Related Work

Loop unrolling belongs to the class of compiler optimizations that has been extensively studied for ACET reduction. In [SK04], generalized loop unrolling methods were presented that allow the optimization of *badly-structured* loops which are transformed into *well-structured* loops before being optimized. A combination of loop unrolling and automatic scheduling by the compiler was discussed in [LH95]. The positive effects of loop unrolling concerning an increased instruction-level parallelism were exploited for instruction scheduling in [MCG+92].

One of the central questions for loop unrolling is how to compute an appropriate unrolling factor. It has been shown that the consideration of I-cache constraints and register pressure is significant for that computation. The problems of automatically selecting an appropriate unrolling factor were addressed in [Sar01]. In addition to the consideration of I-caches and register availability, the authors used a technique called *unroll-and-jam* to produce compact code for nested loops. [KKF97] described an unrolling heuristic taking information about data dependencies, reuse, and machine resources into account. Heydemann [HBK+01] presented an integer linear programming based approach to compute the unrolling factor. However, his approach exhibits strong limitations. For example, no conditions in the unrolled loops are allowed, I-cache interference is not respected, and only unrolling factors being a power of two are considered.

In addition, the selection of an appropriate unrolling factor essentially depends on the knowledge of loop iteration counts. Many works implicitly assume that loop iteration counts are given and do not address their determination. In [DJ01], a survey of different aspects of unrolling was given. One of the main conclusions is that missing loop bounds considerably decrease optimization potential. They conducted profiling to gather this crucial parameter. However, as described in Sect. 4.5, profiling information might not be reliable. Moreover, profiling is expensive and thus often not applicable. To overcome this dilemma, WCC's static loop analysis is applied to compute safe and input-invariant loop bounds that are valid for all input data.

The only work that considers loop unrolling in the context of WCET minimization is presented by Zhao et al. in [ZKW+05]. However, this work differs in several aspects from the approach presented in this book. Most important, Zhao did not develop a WCET-aware loop unrolling but applied standard ACET unrolling and studied its impact on the WCET. Moreover, he did not exploit worst-case iteration counts of loops for an aggressive WCET optimization, but unrolled each loop by a constant factor of two. Also, his unrolling is applied at the assembly level, and his target architecture has no caches, making an evaluation of cache effects impossible.

4.5.3 Standard Loop Unrolling

The unrolling factor u defines the number of replications of the loop body. A review of previous literature has shown that there is no standardized definition for the amount a loop is unrolled. In this book, the term *unrolling factor* is used as follows:

Definition 4.7 (Unrolling factor) The *unrolling factor* u specifies how often a loop body is replicated. If the body of the original loop is replicated n times, then the unrolling factor is $u = n + 1$, i.e., the unrolled loop will contain $n + 1$ loop bodies.

If the loop iteration count of a rolled loop does not correspond to an integral multiple of the unrolling factor u, then either additional exit conditions must be added into the unrolled loop body or some *left-over* iterations must be separately handled by an additional loop (*remainder loop*).

```
                                           for ( i=1;  i<5;  i+=j ) {
                                               j = 0;
                                               x[i] = p * x[i-1];
   for ( int i=1; i<5; ++i ) {                 ++j;
       x[i] = p * x[i-1];          →           if ( i+j>=5 ) break;
   }                                           x[i+1] = p * x[i];
                                               ++j;
                                               if ( i+j>=5 ) break;
                                               x[i+2] = p * x[i+1];
                                               ++j;
                                           }
```

Fig. 4.19 Example for loop unrolling

Loop unrolling can be performed at both the assembly or source code level. WCC's loop unrolling is performed at source code level in the ICD-C frontend. Figure 4.19 demonstrates a possible implementation of unrolling of an ANSI C loop using an unrolling factor of 3. The example points out the difficulties emerging when an unqualified factor is selected. To control the loop iterations, a costly test of i is required.

Unrolling a loop positively affects the program performance for many reasons. The most important are [BGS94, Mow94, Muc97]:

- Reduced loop overhead comprising the increment-and-test instructions
- Reduced number of jumps back to the loop's entry might improve pipeline behavior
- Fundamental code transformation for superscalar processors since unrolling makes instruction level parallelism explicit
- Improved spatial locality leading to an improved I-cache performance
- Enabled application of other optimizations in unrolled loops

Since optimizations following unrolling might benefit from the unrolled code, standard literature proposes to execute unrolling as early as possible in an optimization sequence [Muc97]. For this reason, loop unrolling within WCC is performed in the ICD-C frontend.

Despite the large number of positive effects, it has been observed that loop unrolling can also have an adverse impact on the program's performance when the optimization is not applied elaborately. Since unrolling is a code-expanding transformation, an aggressive loop unrolling can overflow the I-cache leading to additional capacity cache misses that did not arise for the rolled loop [DJ01]. An excessive loop unrolling can also lead to additional spill code when the register working set in the unrolled loop exceeds the number of available registers [CK94]. Furthermore, remainder loops should be introduced with caution. They increase the code size but only a small fraction of the program execution time is spent in this code [Sar01]. It should be also noted that unrolling can increase the compilation time resulting from more code that has to be processed by subsequent optimizations. However, this issue is of less importance for embedded systems compilers which primarily focus on performance improvements accepting longer compilation times.

To sum up, a sophisticated loop unroller should take I-cache constraints and the register pressure into account as well as avoid jumps in order not to cancel the optimization benefits when loops are unrolled.

4.5.4 WCET-Aware Loop Unrolling

The central question for loop unrolling is which unrolling factor should be used for each loop. Its computation depends on several parameters:

1. Loop iteration counts of each loop
2. I-cache and free program memory constraints
3. Approximation of spill code generation

In the following sections, these parameters will be discussed in more detail. It will be shown how these parameters are involved in WCC's novel heuristics for an effective WCET-aware loop unrolling.

4.5.4.1 Worst-Case Loop Iteration Counts

The determination of the unrolling factor requires the knowledge about the loop iteration counts at compile time. This information can be provided either by profiling or by a static loop analysis. Profiling is the most common approach. However, the use of profiling data limits the scope of application since it is difficult to obtain reasonable profiles and the generation of this data might have high space and time requirements [PSKL08].

In contrast, a sophisticated static loop analysis is often not available in most compilers. Due to the lack of techniques to compute the iteration counts, most compilers only use a small and constant unrolling factor (typically 2 or 4) [HBK+01, ZKW+05] which avoids negative side-effects due to extensive unrolling but also does not sufficiently exploit the optimization potential.

Some compilers provide a loop analysis that is only able to analyze simple, well-structured loops which are rarely found in real-life applications. Thus, many loops remain unoptimized. WCC's loop unrolling is combined with its integrated polyhedral loop analyzer (see Sect. 3.6) that is able to detect and analyze most loops found in today's embedded systems applications. To the best knowledge of the author, this book provides the first study that evaluates the effectiveness of a sophisticated loop analysis in the context of loop unrolling.

As the discussion on procedure cloning indicated in Sect. 4.3, many real-life applications include data-dependent loops that have *variable* loop bounds, i.e., loops that have different iteration counts depending on the context in which a function is invoked. Many simple loop analyzers found in today's compilers are able to handle only trivial counting loops and do no support data-dependent loops. Thus, such compilers can not explore the full optimization potential.

In contrast, WCC's loop analysis supports data-dependent loops. The analyzer considers all possible inputs, thus the dynamic behavior of each loop is safely approximated. Exploiting the computed worst-case (context-sensitive) iteration counts, an appropriate unrolling factor can be determined that circumvents adverse test conditions or an introduction of remainder loops. This is illustrated in the following.

Example 4.3 Given the following code example with a data-dependent loop in function **foo**:

```
void foo( int n ) {                 int main( void ) {
   for( int i = 0; i < n; ++i ) {      foo( 6 );
      loop body                        foo( 12 );
   }                                    foo( 24 ); ...
}                                   }
```

WCC's loop analyzer would detect the context-sensitive iterations of 6, 12, and 24 for the **for**-loop. In order to avoid costly handling of left-over iterations, the unrolling factor u is chosen such that the iteration counts correspond to an integral multiple of u. Possible unrolling factors are $u = \{2, 3, 6\}$.

For the sake of completeness, it should be noted that it is also possible to unroll loops when the loop counts are not known at compile time. This approach requires the insertion of additional code calculating the iteration counts during the program execution. However, as shown in [DJ01], the cost of calculating the loop iteration counts does not always amortize and potentially leads to a performance degradation. Moreover, this approach prevents the determination of an unrolling factor that takes I-cache and spill code constraints into account. Therefore, this class of loops is not considered within WCC.

4.5.4.2 I-Cache and Memory Constraints

After the loop analysis, an appropriate unrolling factor u has to be calculated based on the determined loop iteration counts. For loops with constant loop bounds i, a complete unrolling is practically feasible, by setting $u = i$. However, such a strategy is often a bad choice mainly due to these two reasons:

- The size of the unrolled loop might exceed the available program memory which is in particular crucial for embedded systems with restricted memory resources.
- A too heavily unrolled loop body may lead to I-cache thrashing.

In order to circumvent such adverse effects, the optimization must be able to estimate the code size of the finally unrolled loop. This knowledge allows the calculation of an unrolling factor that restricts the loop size increase to a given I-cache capacity.

The determination of a precise loop size requires knowledge about the involved assembly instructions. On the one hand, if loop unrolling is applied at assembly level, the loop size can be easily extracted and used for the I-cache constraints. However, due to the late application of the optimization, other optimizations performed previously cannot benefit from the unrolled code. On the other hand, unrolling loops at the source code level offers optimization opportunities for a larger set of following optimizations, but a precise estimation of code size at source code level is usually not possible. Thus, both solutions are not fully satisfactory.

Exploiting Back-Annotation   WCC's loop unrolling takes advantage of both code abstraction levels. Unrolling takes place at the source code level providing optimization potential for subsequent source code and assembly level optimizations. To estimate the assembly size of the loop, the compiler transforms each loop into assembly code. At this level, the size of the loop header and loop body can be easily determined in bytes. Using WCC's back-annotation (cf. Sect. 3.7), this information is imported back into the compiler frontend and made available to the source code optimizer. Since unrolling is placed in the applied sequence of WCC's source code optimizations after all code-expanding transformations, e.g., *function inlining* or *loop unswitching*, a precise estimation of the loop size is possible. This data is involved in the determination of the unrolling factor to avoid I-cache thrashing.

To decide how extensively loops can be unrolled without violating memory constraints, the loop optimizer extracts various information from WCC's memory hierarchy specification that was introduced on p. 36. The first derived data concerns the I-cache capacity of the target architecture. Taking this parameter into account, an appropriate unrolling factor can be chosen that limits the size of the unrolled loop to the given cache size. Moreover, WCC's optimizer is provided with detailed information about the memory usage of the program under analysis. The compiler constructs a memory usage model of the physical memories available in the underlying target architecture. During code generation and code modification, this model is kept up-to-date such that valid information about memory usage can be extracted at any time.

Together with the I-cache constraints, the amount of free space in the program memory is obtained during the calculation of the unrolling factor in order to find a value that does not exceed the available program memory when the loop is unrolled. In addition, the user can parameterize the memory usage considered during loop unrolling. For example, the user might want to use only half of the I-cache, or just allow loop unrolling to consume 60% of the free memory. With this flexible handling of the memory model, WCC's loop unrolling can be effectively tailored towards particular memory requirements that are often imposed on embedded systems.

The flexible memory management makes this code transformation retargetable since the exchange of the target platform requires exclusively the adjustment of the compiler's memory hierarchy specification.

The exploitation of the back-annotation overcomes the common problem concerning the code size estimation at source code level. In [DJ01], the authors criticized ad-hoc approaches that use a conversion factor to estimate the number of machine-language instructions produced for a source code line. In contrast, WCC provides precise size information that can be exploited as a reliable cost function.

4.5.4.3 Prediction of Unrolling Effects

Besides I-cache overflows, unrolling can lead to the generation of additional spill code which may significantly degrade the worst-case performance of the program (cf. Sect. 3.3.5 on p. 31). These load and store instructions are expensive and should be avoided, otherwise they may cancel the unrolling benefits and even yield a performance degradation.

A possible approach to avoid negative effects due to spilling was presented in the course of the discussion about superblock optimizations in Sect. 4.4. Whenever the application of the optimizations yielded a degraded WCET estimation, the transformations were completely undone. However, this rollback mechanism is not sufficient for loop unrolling. To enable an effective unrolling, the compiler should be capable of predicting the amount of expected spill code depending on the employed unrolling factor. This way, the most promising factor can be computed.

To predict the effects of loop unrolling at source code level on the register pressure, Sarkar [Sar01] proposed an approach that tries to approximate the amount of spill code based on the maximal number of simultaneously live fixed- and floating-point values in the unrolled loop. However, this approach has two main limitations. First, it is inflexible since its applicability highly depends on the involved register allocator. If the register allocator is modified or even exchanged by a register allocator pursuing another allocation strategy, then the spill code approximation might fail. Second, as stated by the author, their approximation is conservative and may unnecessarily limit the amount of permitted unrolling.

Such ad-hoc approaches are required when information from the compiler back-end is not present. WCC uses a more realistic prediction of unrolling effects on the spill code exploiting the back-annotation. The prediction is based on a comparison between original loops and their unrolled version. During the evaluation, a copy $\hat{P}$ of the original program $\mathcal{P}$ under analysis is created. In a second step, loops in $\hat{P}$ are virtually unrolled and a comparison between the original and unrolled loops reveals if additional spill code was generated.

In detail, for each loop L in $\hat{P}$, all context-sensitive loop iteration counts derived from WCC's loop analysis are considered. Based on the set S of possible iteration counts i, the *smallest common prime factor (SCPF)* for each loop L is determined:

Definition 4.8 (Smallest common prime factor) Let GCD_L be the greatest common divisor $\forall i \in S$ of a loop L. The *smallest common prime factor* $SCPF_L$ is the smallest prime factor of GCD_L, if $GCD_L > 1$. Otherwise, $SCPF_L$ is 1.

Example 4.4 Given the code from Example 4.3. For the set of context-sensitive loop iteration counts of the **for**-loop consisting of the elements $\{6, 12, 24\}$, the greatest common divisor $GCD_L = 6$. Thus, the smallest common prime factor for this loop is $SCPF_L = 2$.

The utilization of the *SCPF* for the prediction of unrolling effects is motivated by WCC's unrolling strategy that tries to avoid additional conditional jump statements in the finally unrolled loops. Jumps are required when loops are unrolled with unrolling factors that do not evenly divide the number of iterations. Conditional jumps have several negative effects on the program performance and a static WCET analysis:

- Jumps degrade pipeline behavior since additional control pipeline hazards are introduced.
- High penalty cycles may be encountered if jumps are mispredicted by the processor's branch prediction.
- Results from static program analyses may become imprecise since computations at jumps require to be merged.

Thus, the optimization unrolls loops such that additional jumps are not required.

To be consistent with the final unrolling strategy, jumps in the virtually unrolled loops in $\hat{P}$ have to be avoided. This could be achieved by employing GCD_L as unrolling factor. However, this factor is often not applicable due to memory resource constraints. Some loops have a large GCD of their context-sensitive execution counts. If the loops were unrolled by this factor, they would possibly exceed the available program memory which is often strongly restricted when multiple tasks reside in the same memory. Using *SCPF* for the evaluation, jumps in the unrolled loops and memory overflows are typically avoided since the *SCPF* is usually significantly smaller than the *GCD*.

The impact of unrolling on the spill code generation is predicted as follows:

1. Create a copy $\hat{P}$ of original program P
2. Virtually unroll each loop $\hat{L} \in \hat{P}$ with $u = SCPF$
3. Generate assembly code and perform back-annotation for P and $\hat{P}$
4. Compare each $\hat{L}$ with respective L

To decide whether additional spill code was generated after unrolling, the *spill code ratio* ψ_u^L for each loop L and a given unrolling factor u is computed:

Definition 4.9 (Spill code ratio)

$$\psi_u^L = \frac{\sum \text{spilling instructions} \in L}{\text{unrolling factor } u}$$

It was empirically determined that ψ_u^L is a reliable indicator for an augmented spill code generation: if ψ_u^L does not increase for $u = SCPF$ compared to the original loop before unrolling, i.e., no additional spilling instructions were generated, then it is very likely that no extra spill code will be introduced for $u > SCPF$.

WCC's spill code estimation is not only more accurate but also more flexible compared to the approach described by Sarkar [Sar01]. Unlike Sarkar's estimation, WCC's estimation of the spill code is not carried out at the code provided to the loop unroller. Rather, it is based on the back-annotated spilling-related data of the

---

Algorithm 4.3 Algorithm for determination of final unrolling factor

---

Input: L, *freePMem*, *cacheSize*
Output: u^L_{final}

 1: $sizeHeader \leftarrow backannotate(L)$
 2: $sizeBody \leftarrow backannotate(L)$
 3: $set\langle int \rangle\ iterations \leftarrow loopAnalysis(L)$
 4: **for** $i = \gcd(iterations)$ **to** 1 **do**
 5:    $sizeUnrolled \leftarrow sizeHeader + i \cdot sizeBody$
 6:    **if** $(i - 1) \cdot sizeBody \leq freePMem$ **and**
      $sizeUnrolled \leq cacheSize$ **and**
      $\gcd(iterations)\ \bmod\ i == 0$ **then**
 7:       **return** i
 8:    **end if**
 9: **end for**
10: **return** 1

(possibly further optimized) code that is provided to the register allocator. Hence, optimizations following unrolling are allowed and their effects on spilling is respected.

Similar to the spill code estimation, virtual unrolling with $u = SCPF$ is applied by WCC to predict the effects of loop unrolling on the loops' WCET estimates. This data is back-annotated into ICD-C and made available to the optimizer.

4.5.4.4 Determination of Final Unrolling Factor

The virtual unrolling of loops serves the purpose of predicting the effects of unrolling on the WCET estimation, code size, and spilling when loops are carefully unrolled using *SCPF*. However, this unrolling factor is deliberately chosen small and does not enable a full exploitation of the optimization potential. To this end, a final unrolling factor $u^L_{final} \geq SCPF$ for each loop L that maximally replicates the loop bodies has to be determined.

The determination of the final unrolling factor is shown in Algorithm 4.3. Besides a loop unrolling candidate L, the algorithm requires information about the status of the free program memory (*freePMem*). This data is dynamically computed based on the memory hierarchy specification and the current state of $\mathcal{P}$. A further input parameter is the cache size of the target architecture (*cacheSize*).

After collecting information about the code size (lines 1–2), the algorithm determines context-sensitive iteration counts of loop L (line 4). Next, the algorithm explores possible unrolling factors (lines 4–9) and returns the maximal factor that does not violate memory constraints (i.e., free program memory and cache size) and avoids control jumps for left-over iterations (lines 5–6). Since unrolling is applied after all code-expanding optimizations, the free program memory can be employed for the unrolled code.

Algorithm 4.4 Algorithm for WCET-aware loop unrolling

Input: $\mathcal{P}$
Output: *unrolled* $\mathcal{P}$

1: $\hat{\mathcal{P}} \leftarrow \mathcal{P}$
2: $set\langle Loop\rangle\ loops_{SCPF} \leftarrow \hat{\mathcal{P}}$
3: **for all** $\hat{L} \in loops_{SCPF}$ **do**
4:     $unroll_{SCPF}(\hat{L})$
5: **end for**
6: $evaluateUnrolling_{SCPF}(loops_{SCPF})$
7:
8: $set\langle Loop\rangle\ loops_{final} \leftarrow \mathcal{P}$
9: $removeSpillingLoops(loops_{final})$
10: $sortByProfit(loops_{final})$
11: **for all** $L \in loops_{final}$ **do**
12:     $unroll_{final}(L)$
13: **end for**
14: **return** $\mathcal{P}$

4.5.4.5 WCET-Aware Unrolling Heuristics

Algorithm 4.4 outlines WCC's final WCET-aware loop unrolling. The algorithm operates in two phases. In the first phase (lines 1–6), virtual unrolling is conducted by unrolling all loops from the copied program $\hat{\mathcal{P}}$ using *SCPF*. These loops are evaluated (line 6) by collecting information about the impact of unrolling w.r.t. the WCET estimation, the code size, and the spilling behavior.

This data is exploited in the second phase (lines 8–13). Each loop L of the original program $\mathcal{P}$, for which its counterparts $\hat{L}$ exhibited an increased spill code ratio ψ_u^L, is omitted from unrolling since adverse effects of spilling turned out to outbalance the benefits of unrolling (line 9). In the next step, loops are sorted by their unrolling profit Φ_{unroll}^L (line 10):

Definition 4.10 (Unrolling profit) Let $WCET_L$, $WCET_{\hat{L}}$ be the WCET and CS_L, $CS_{\hat{L}}$ the code size of the original and the virtually unrolled loop using the SCPF unrolling factor u_{SCPF}^L, respectively. Moreover, let u_{final}^L be the final unrolling factor computed by Algorithm 4.3. The *unrolling profit* Φ_{unroll}^L is computed as follows:

$$\Phi_{unroll}^L = \frac{WCET_L - WCET_{\hat{L}}}{CS_{\hat{L}} - CS_L} \cdot \frac{u_{final}^L}{u_{SCPF}^L}$$

The unrolling profit represents the expected benefit concerning the WCET estimates and code size when L is unrolled with the final unrolling factor. It is used to define the unrolling order, that is, loops with the highest profit are optimized first using the unrolling factor computed by Algorithm 4.3 (lines 11–13). Since loops with a higher profit are preferred, the free program memory is primarily utilized for unrolling of loops that promise the highest decrease of the WCET estimation with a simultaneously low code expansion. Loops with a negative profit are excluded since they will likely degrade the worst-case performance. Hence, this strategy is best suited for embedded real-time systems with restricted resources.

The optimization does not explicitly consider WCEP switches since typically all loops in a program lie on the worst-case execution path. Thus, the high overhead of considering potential path switches which usually entails repetitive runs of the expensive WCET analysis does not pay off. Yet, loops not part of the WCEP do not contribute to the WCET estimation and should not be transformed. This issue is respected during optimization by omitting loops with a back-annotated WCET of 0 cycles.

The profit calculation serves as an estimation of unrolling effects before the actual unrolling. This option is missing in most compilers, thus loops are often unrolled that decrease program performance. Using WCC's profit calculation, it can be detected in advance if a loop should be unrolled or if its unrolling will likely decrease the worst-case behavior. The benefits of WCC's unrolling prediction are demonstrated in the following by results achieved on real-life benchmarks.

4.5.5 *Experimental Results for WCET-Aware Loop Unrolling*

This section evaluates the impact of WCC's WCET-aware unrolling on the WCET estimates of real-life benchmarks. In total, 45 benchmarks from the test suites DSPstone [ZVS+94], MRTC [MWRG10], MediaBench [LPMS97], and UTDSP [UTD10] were involved in the experiments. The code sizes of the considered benchmarks range from 302 bytes up to 14 kByte with an average code size of 1.9 kByte per benchmark. The number of innermost loops considered for loop unrolling ranges between 1 and 15, depending on the benchmark complexity.

The TriCore TC1796 processor, which is the supported target architecture within the WCC framework, is equipped with a 16 kByte I-cache and 2 MByte program Flash that was employed for the following experiments. The I-cache size can be virtually modified for the WCET analysis to evaluate different cache sizes. The Tri-Core processor does not have a data cache. However, unrolling marginally changes memory reference patterns, thus D-cache misses can be ignored for unrolling factor selection [HBK+01].

4.5.5.1 WCET

Figure 4.20 shows the results of unrolling on the estimated WCET. The dark bars represent the average WCET reduction for all 45 benchmarks when the code is compiled with the highest optimization level *O3* including WCET-aware unrolling w.r.t. to the code that was generated using the highest optimization level and disabled unrolling. For the experiments, the I-cache size was modified between 512 bytes

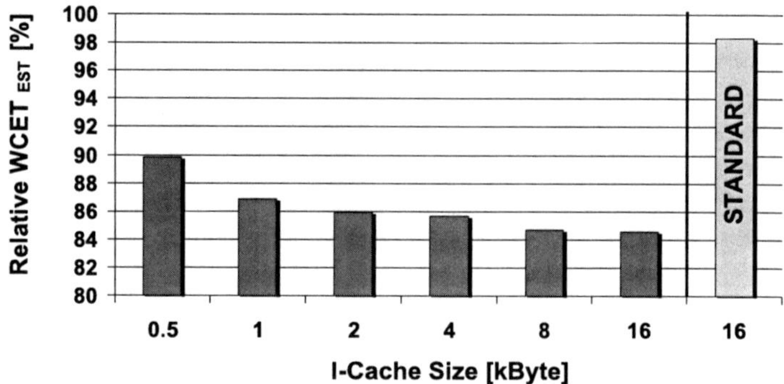

Fig. 4.20 Relative WCET estimates for WCET-aware loop unrolling

and 16 kByte in order to demonstrate the impact of unrolling on different cache architectures.

As can be seen, the WCET can be decreased from 10.2% for the smallest cache capacity (512 bytes) up to a WCET decrease of 15.4% for the largest cache (16 kByte). The reason for this increase is obvious. Using larger caches, WCC is provided with more optimization opportunities since extensive unrolling is less frequently limited by the I-cache constraints. It can also be seen that for larger caches the WCET reduction becomes smaller. The reason are smaller benchmarks that can be already fully optimized for modest cache sizes. Relaxing I-cache constraints due to larger cache capacities does not result in any further benefit.

The results for WCET-aware unrolling were compared with standard ACET loop unrolling which restricts the size of the unrolled loops to 50 expressions. For 16 kByte I-cache, the standard approach achieved an average WCET reduction of 1.7% (see light bar on the right-hand side of the figure). Hence, WCET-aware loop unrolling outperforms the standard optimizations by 13.7%.

An interesting question is whether the WCET reductions can be attributed to the additional information generated by the static loop analysis, or whether the novel, WCET-aware unrolling heuristics are the main source for the decreased WCET. Figure 4.21 answers this question by examining the benefits of WCC's different unrolling strategies.

The results show the estimated WCET for three different strategies using a 2 kByte I-cache to capture cache effects. 100% corresponds to the WCET of the benchmarks compiled with the highest optimization level and disabled unrolling. For the sake of readability, results on a subset of 19 representative benchmarks are shown. As can be seen, applying WCC's standard ACET loop unrolling (first bar labeled with *Standard LU*), which uses a simple context-insensitive loop analysis and restricts the size of the finally unrolled loop to 50 expressions, has minimal positive effects on the WCET. Integrating WCC's sophisticated context-sensitive loop analysis (second bar labeled with *Standard LU + LA*) into standard unrolling, slightly improves the average WCET estimation by 2.9%.

Notably improved results are achieved when loop unrolling is extended by WCC's novel WCET-aware heuristics represented by the last bar per benchmark.

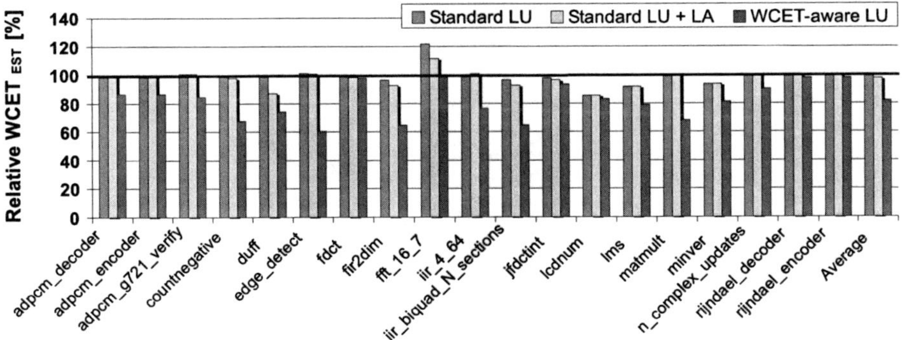

Fig. 4.21 Comparison of unrolling strategies

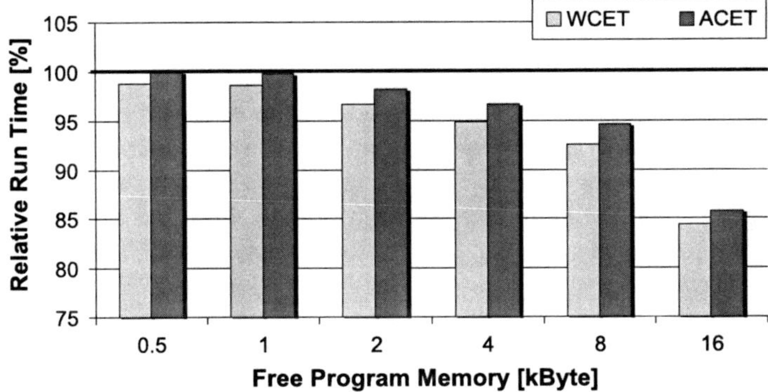

Fig. 4.22 Impact of program size unrolling heuristic on *ndes*

WCET reductions of up to 39.5% are observed. These high WCET reductions achieved e.g., for the benchmarks *edge_detect* and *fir2dim* are the result of a complete unrolling of some loops. The generated straight line code can be effectively improved by succeeding optimizations and all jumps to loop entries are entirely eliminated.

Based on a review of the optimized code, it can not be said in general which unrolling effects are most beneficial for the performance of the code. For some benchmarks, like *countnegative*, the reduced loop overhead was the key factor, while other benchmarks availed of enabled optimization potential after unrolling.

4.5.5.2 ACET

To indicate the practical use of the program memory heuristics, a system environment is simulated for which program memory is restricted. This situation is common for embedded systems that execute multiple tasks residing in the same memory. Figure 4.22 shows the results for the WCET and ACET when WCET-aware unrolling

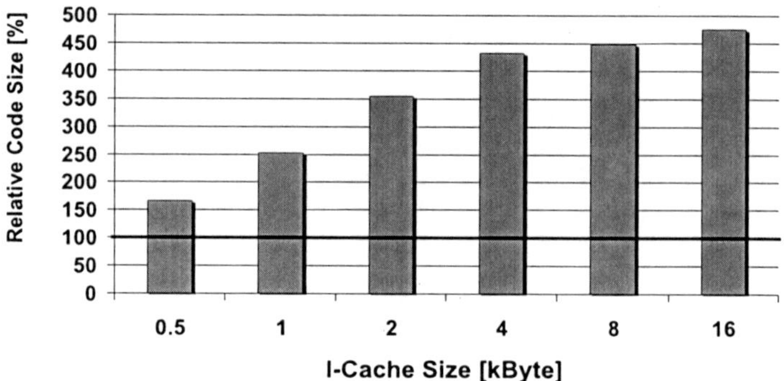

Fig. 4.23 Relative code size for WCET-aware loop unrolling

is applied together with *O3* for the example MRTC benchmark *ndes*. 100% correspond again to the program's run times for *O3* and disabled unrolling. For the determination of the WCET estimation and the ACET (using a cycle-true simulator), the standard I-cache of the TC1796 with a capacity of 16 kByte is utilized. First, it can be seen that with increasing program memory, the reduction of the benchmark's WCET is also increasing. This is expected as with more program memory, unrolling can be applied more aggressively. Second, from the results it can be inferred that the presented optimization is tailored towards WCET minimization. For all memory sizes, a higher reduction of the WCET than the ACET is achieved. One reason for this behavior is that WCC begins to unroll those loops that promise the highest WCET reduction. Since the program memory is restricted, not all loops can be unrolled and the transformed loops most beneficial for WCET reduction need not correspond to the most beneficial loops for ACET reduction.

4.5.5.3 Code Size

Figure 4.23 shows the impact of WCC's WCET unrolling on the average code size w.r.t. to the average code size of the benchmarks with disabled loop unrolling that corresponds to 100%. Six different I-cache configurations ranging from 512 bytes up to 16 kByte are considered. As expected, the code size increase becomes larger with an increasing cache size. This is due to the growing potential of unrolling since loops can be unrolled more aggressively for larger caches without exceeding their capacity.

It should be noted that the average code size increase is mainly reflecting small benchmarks, while the code size increase for larger benchmarks is modest. For example, the highest increase was found for the small DSPstone benchmark *matrix1* with 1261% for the 16 kByte cache, resulting in a final code size of 4114 bytes. However, this code size is still fully acceptable for modern embedded systems. In contrast, the code size increase for the largest benchmark, *rijndael_encoder* from MediaBench, exhibited only a maximal increase of 6.4%. This is also a typical scenario for other larger benchmarks.

Concerning the absolute code sizes, it can be concluded that WCC's WCET-aware unrolling produces enlarged code of a size that is acceptable for modern embedded systems. If, however, stringent memory constraints have to be met, the optimizer can be provided with user-defined parameters that limit the code expansion.

4.5.5.4 Optimization Run Time

Finally, the optimization run time of WCC's WCET-aware loop unrolling was measured on an Intel Xeon 2.4 GHz system with 8 GB RAM. Compilation of all 45 benchmarks including a single WCET analysis took 29 minutes when all benchmarks are compiled with *O3* including the standard loop unrolling with a simple loop analysis. In contrast, the compilation time for the WCET-aware unrolling took 390 minutes in total, using a 16 kByte I-cache.

This increase mainly results from an additional WCET analysis (used for profit computation), the static loop analysis, and the standard optimizations. Although the compilation times are depending on the complexity and unrolling factor of the benchmark, it can be observed that the major fraction of the additional time was consumed by standard optimizations which have to analyze (in some cases significantly) enlarged basic blocks.

However, as the main objective of embedded systems compilers is performance maximization even at the cost of a longer compilation time, the observed compilation times can be considered as fully acceptable.

The techniques presented in this section have been published in [LM09].

4.6 Accelerating Optimization by the Invariant Path

Compared to traditional compiler optimizations, the previous sections have shown that WCET-aware optimizations are more demanding. To accomplish a systematic improvement of the worst-case behavior, a detailed timing notion of the application under analysis is essential. This data is exploited by a compiler to identify portions of the code that contribute to the overall program WCET. From the perspective of a compiler operating on a control flow graph, these crucial portions of the code are part of the WCEP.

However, for an effective WCET reduction it is not sufficient to determine the WCEP once and to apply all optimization steps on this initial path. This is due to WCEP switches that may emerge if code is optimized such that another path becomes the longest path. To cope with this problem, the WCEP has to be permanently updated during optimization. Usually, this update is accomplished by pessimistically conducting a WCET analysis after each code transformation. This exhaustive WCEP update is time-consuming and may even turn an optimization into an infeasible problem for practical use.

In many situations, however, a code modification does not yield a WCEP switch, making a WCET analysis superfluous. WCC identifies these situations and exposes those parts of the WCEP that are not prone to a path switch. The respective sub-paths of the CFG are denoted as *invariant path* and the exploitation of this knowledge may significantly cut the run time of WCET-aware code optimizations.

The rest of this section is organized as follows. In Sect. 4.6.1, the stability of the WCEP w.r.t. code modifications is motivated by an example. An overview of related work is provided in Sect. 4.6.2. Different path scenarios found in a CFG and their invariance to a WCEP switch inducing the definition of the invariant path are discussed in Sect. 4.6.3. The construction of the invariant path and its ratio in the code of real-life benchmarks is presented in Sect. 4.6.4 and Sect. 4.6.5, respectively. To demonstrate the practical use of this new concept, the knowledge about the invariant path is exploited in order to accelerate the novel optimization WCET-aware loop unswitching. The optimization is discussed in Sect. 4.6.6, followed by experimental results in Sect. 4.6.7.

4.6.1 Motivating Example

Branches in the control flow graph are the source of WCEP switches. They split the CFG into n mutually exclusive paths $\pi_1, \ldots, \pi_n$ and an initial WCEP following path π_k, with $1 \leq k \leq n$, may switch to path π_l, with $1 \leq l \leq n \wedge l \neq k$, when an optimization shortened the length of π_k such that π_l becomes the longest path afterwards.

However, the WCEP switch does not necessarily arise for all branches. Context-sensitive WCET analyses distinguish between different invocations of functions and between different loop iterations. Consequently, branches, which can be statically evaluated during analysis, may show different outcomes depending on the execution context. In contrast to timing analyzers that aim at a high precision, compilers usually provide a context-insensitive view of the dynamic behavior of the program. That way, they manage the high analysis complexity at the cost of precision which is in general not a compiler's key objective.

To this end, the import of context-sensitive analysis results into a compiler requires a translation of these results. To warrant correctness, a safe approximation of the context-sensitive results is required: for each CFG edge, a union of all context-sensitive WCEP results is computed. This union represents the context-insensitive WCEP information of each CFG edge. Referring to branches, each mutually exclusive path that was taken during the static WCET analysis in some execution context, is—from the view of a compiler—an optimization candidate.

Since the semantics of branch conditions does provably not change during program optimization, each initial static evaluation of a branch condition will remain valid. Thus, each mutually exclusive path starting at a statically evaluated branch that was part of the WCEP before program optimization will also lie on the WCEP after any code modification. As a consequence, WCEP switches to other mutually exclusive paths are infeasible. Compilers can exploit this knowledge to avoid redundant path updates by time-consuming WCET analyses.

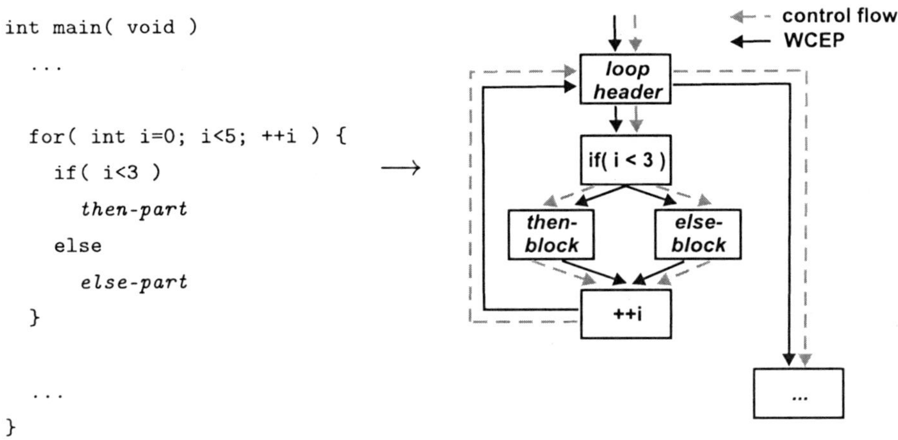

Fig. 4.24 Example for invariance of the worst-case execution path

Figure 4.24 illustrates the invariance of the WCEP for a simple example program written in ANSI C. The code on the left-hand side of the figure contains a `for`-loop which is executed five times. A static WCET analysis can compute that the loop's `if`-condition evaluates to *true* for the first three iterations, while *false* for the remaining loop iterations. The right-hand side of the figure shows the corresponding CFG with the incorporated results of the WCET analysis. The dashed edges mark the regular control flow while the solid edges represent the WCEP. As can be seen, the WCEP covers both paths of the `if`-statement in different contexts. In this situation, the optimization of neither the *then-* nor the *else*-part can lead to a WCEP switch in this branch. Hence, shortening of both mutually exclusive paths will effectively reduce the program's WCET. A WCEP recomputation after the modification of these paths is not required since the risk of operating on an outdated WCEP does not exist.

This section discussed one branch scenario which is not prone to a WCEP switch. In the terminology of the WCC framework, such stable parts of the WCEP are called invariant path. The challenge is to classify all possible branches within a program as being part of the invariant path or not. Section 4.6.3 provides a thorough discussion about this kind of classification.

4.6.2 Related Work

The WCEP switch is an inherent problem of WCET-aware optimization. Some related works do not consider switching WCEPs, thus there is no guarantee that a WCET reduction with each optimization step is achieved. An example is [PD02] where the authors proposed two algorithms for static I-cache locking. The employed heuristics for cache content selection are based on the WCEP, which is determined before the optimization, while subsequent steps operate on this initial, possibly out-dated WCEP. Another example is D-cache locking for WCET minimization presented in [VLX03] where a WCEP is not considered at all.

Other works enable an effective WCET minimization by regularly recomputing the WCEP after each optimization step via a static WCET analyzer. For example, Falk [Fal09] presented a register allocation which reduces the generation of spill code on the WCEP. The path information is updated after processing of each basic block. In [ZWHM04], a code positioning optimization driven by WCET data was discussed. By rearranging the memory layout of basic blocks, branch penalties along the WCEP are avoided. Each rearrangement is preceded by a WCET analysis.

A trade-off between precision and optimization run time was presented in [Pua06]. The author extend the techniques of [PD02] for I-cache locking by taking the WCEP switch into account. To reduce the optimization time, the WCET data is not updated after each single code transformation but after n steps, where n is a user-defined parameter. As the author showed, an updated WCEP is crucial since experiments with infrequent path updates lead to less effective WCET minimization.

Another elegant approach to cope with the switching path is to formulate the WCET minimization problem as an ILP model. This way, the WCEP and its switches are implicitly considered. The few published works based on this technique comprise the optimal static WCET-aware scratchpad allocation of program code [FK09] and data [SMR+05]. However, the ILP-based approach can not be applied to all optimization problems since the formulation of an ILP model is often too complex or even infeasible for particular problems. Moreover, solving of ILP models is $\mathcal{NP}$-complete, thus it may be not applicable for some problems in practice.

None of these papers explicitly addresses the problem of the invariance of the WCEP. The only paper the author is aware of, which discusses a related topic, was published by Giegerich et al. [GMW81]. The authors discussed the invariance of approximative semantics after program transformations. In detail, they constructed a framework based on abstract program semantics which is invariant to the elimination of dead code, i.e., the abstract transformation rules are constructed such that particular flow information remains valid without the need of their recalculation each time a transformation rule is applied. The paper does not provide any results, thus it can not be judged how beneficial these concepts are for practical use. Moreover, the concepts presented in [GMW81] are not suitable to identify invariant paths since they require an analytical modeling of each program transformation as approximative semantics. For compilers with a large number of supported transformations, this overhead is not feasible.

4.6.3 Invariant Path Paradigm

As motivated previously, branches within a CFG are the locations where a WCEP switch may emerge. The majority of available WCET-aware compiler optimizations cope with this problem in a conservative way—they recompute their WCET data using a static timing analyzer each time code was modified. However, as indicated in Sect. 4.6.1, particular branch scenarios provably exclude WCEP switches between their mutually exclusive paths, removing the conservatism to recompute timing information. In this section, different control flow scenarios found in a CFG are explored to determine whether they are prone to a path switch.

The stability of the WCEP is expressed within WCC by the concept of the invariant path π_{inv}:

Definition 4.11 (Invariant path) The *invariant path* π_{inv} is a sub-path of the WCEP that always remains part of the WCEP independent of the applied code modifications to the program $\mathcal{P}$.

Obviously, any straight-line code that lies on the WCEP and is not part of any mutually exclusive paths is declared as invariant path. The challenging constructs found in a CFG constitute control flow branches.

Modeling of branches in a CFG depends on the concrete semantics of a programming language. High-level languages typically use so-called *selection statements*, such as ANSI C `if-`, `if-else-`, or `switch-` (general form of `if-else`) statements, to model mutually exclusive paths. They perform a conditional execution of the code dependent on a condition expression. Low-level programming languages model mutually exclusive paths using conditional jump instructions. Other types of statements/instructions that alter the program's control flow, like *call* statements, are not considered by the invariant path since they are irrelevant for a WCEP switch.

In the following, possible branch types found in WCC's source language ANSI C are discussed regarding their invariance properties of the WCEP. However, the concepts of the invariant path are generic and can be easily adapted to other programming languages.

4.6.3.1 *IF-THEN* Structure

An *IF-THEN* branching structure is modeled in ANSI C as an `if`-statement which represents a conditional execution. Depending on the conditional expression, either the path through the *then*-part is executed or the mutually exclusive path bypassing the *then*-part is followed. In this scenario, the WCEP either traverses the *then*-part or the *then*-part does not contribute to the WCET.

In terms of the invariant path paradigm, a WCEP that traverses the *then*-part is also part of the invariant path π_{inv} since a modification of the code in the *then*-part can not yield a path switching. This is because the other feasible path of the `if`-statement does not contain any code that might become the new WCEP.

The scenario in which a context-sensitive static analysis computes a WCEP, that traverses both branch paths in different contexts, is depicted in Fig. 4.25. As can be seen, the entire branch structure contributes to the invariant path. This knowledge can be exploited during the WCET-aware optimization. If code in the *then*-part was

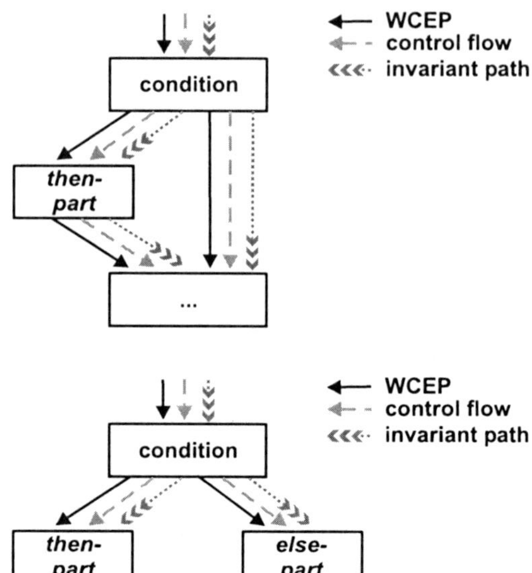

Fig. 4.25 Invariance of *IF-THEN* structure

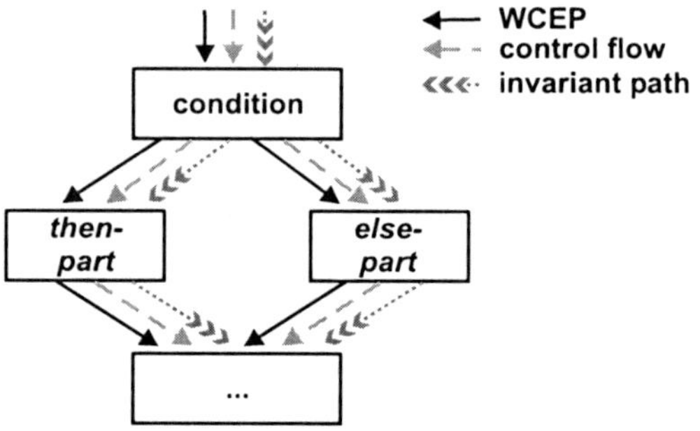

Fig. 4.26 Invariance of *IF-THEN-ELSE* structure

modified in the previous optimization step, there is no need to update the WCEP by an expensive WCET analysis for the next step since a WCEP switch can be excluded.

4.6.3.2 *IF-THEN-ELSE* Structure with Statically Evaluable Condition

For the branch type resulting from an ANSI C `if-else`-statement, the context-sensitive WCET analysis may determine that the WCEP traverses both the *then*- and *else*-part in different execution contexts. This scenario was already discussed in the motivation (cf. Sect. 4.6.1).

For a WCET optimization that must be aware of a valid WCEP, an `if-else`-statement, for which the WCEP traverses both branch paths, is not crucial since a path switch can not emerge. By treating this selection statement in a context-insensitive manner, it is known from the WCET analysis that both blocks always contribute to the program's WCET. Hence, this type of branch scenario can be declared as part of the invariant path as shown in Fig. 4.26. WCET-aware optimizations can operate on both paths and there is no risk that transformations on an irrelevant path are applied. Thus, costly WCEP updates are superfluous.

The other possible scenario is that the static WCET analysis is able to detect that the condition of the `if-else`-statement always evaluates to *true* or *false*. As a consequence, one of the two paths is never executed and can be defined as dead code. Excluding the dead branch path reduces the `if-else`-statement to an `if`-statement, enabling the classification of the remaining path as invariant path.

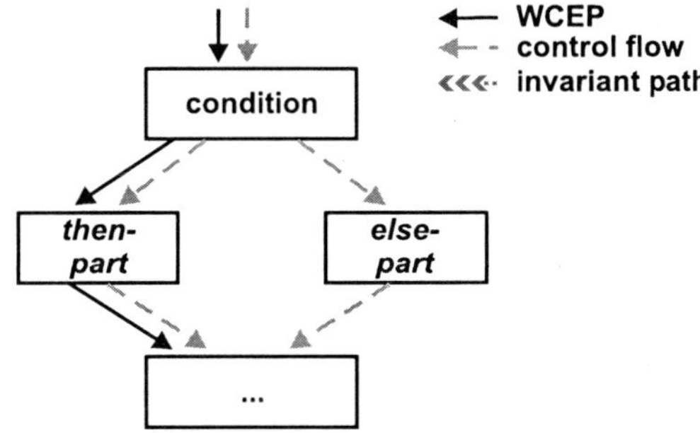

Fig. 4.27 Potential WCEP switch in *IF-THEN-ELSE* structure

According to WCC's terminology, this type of branches as well as branches resulting from `if`-statements are called *good-natured* since they are not prone to a path switch. This is the typical situation found in most applications, thus a large portion of the code can be declared as invariant path.

4.6.3.3 *IF-THEN-ELSE* Structure with Statically Non-evaluable Condition

The last class of selection statements differs from the previous one in terms of the WCEP flow. For this scenario, the WCET analysis assumes that for all execution contexts the WCEP traverses exactly one of the mutually exclusive paths, i.e., either through the *then-* or the *else*-part, while the other path does not contribute to the WCET. An example with an excluded *else*-part is depicted in Fig. 4.27.

This situation occurs if the condition of the `if-else`-statement can not be statically evaluated by the timing analyzer. In this situation, the timing analysis must conservatively assume that the longer of the mutually exclusive paths is the WCEP which is traversed in all contexts. This is the only situation where a WCEP switch can occur. These types of branch statements are not declared as invariant path.

As illustrated in Fig. 4.27, a code transformation may optimize the *then*-part such that the path through the *else*-part becomes the new WCEP. To make sure that the next step of the optimization does not operate on an outdated WCEP, a validation of the path information by a WCET analysis is required.

4.6.4 *Construction of the Invariant Path*

The construction of the invariant path is performed recursively based on the concepts discussed in the previous section. The algorithm begins at the entry point of the program's CFG—typically the first statement or instruction of the *main* function—and traverses the graph in a depth-first search manner.

All blocks lying on the WCEP that are not part of any branch, i.e., straight-line code fragments that are not control dependent on any branch statement, are declared by WCC as invariant path since no mutually exclusive paths for a WCEP switch exist. If branches are detected during CFG traversal, it is decided if they can be classified as invariant path. This decision is exploited in the further search. If a branch is declared as invariant path π_{inv}, all basic blocks of its mutually exclusive paths are also declared as π_{inv} as long as no nested branch during the traversal is reached for which a new classification is conducted. If function calls are encountered on the WCEP declared as invariant path, their bodies are analogously analyzed. This way, all basic blocks of the program's CFG are visited and possibly classified as invariant path. Within the WCC framework, the invariant path construction is carried out on the ICD-C IR (cf. Sect. 3.3) by marking basic blocks that are part of the invariant path. Technically, this is solved by attaching ICD-C persistent objects, which hold the invariance information, to the respective blocks.

4.6.5 Invariant Path Ratio

The concepts of the invariant path are generic (not restricted to any programming language) and can be combined with most WCET optimizations that are aware of the WCEP. Sub-graphs of the WCEP which are declared as invariant path can be modified without invalidating the WCEP information. To get an impression of how sensitive benchmarks are to WCEP switches, the fraction of the code that is part of the invariant path was computed. The computation was performed on 42 benchmarks from the MRTC and MediaBench Benchmark suites.

The stability of the WCEP can be expressed by the *static* and *dynamic invariant path ratio* of the code. Let $IP(b)$ indicate whether a basic block b contributes to the invariant path:

$$IP(b) = \begin{cases} 1, & \text{if block } b \text{ lies on the invariant path,} \\ 0, & \text{otherwise.} \end{cases}$$

The static invariant path ratio rat_{stat} is defined as:

Definition 4.12 (Static invariant path ratio) The *static invariant path ratio rat_{stat}* represents the fraction of basic blocks b of program $\mathcal{P}$ that contribute to the invariant path:

$$rat_{stat} = \frac{\sum_{b \in \mathcal{P}} IP(b)}{|\mathcal{P}|}$$

with $|\mathcal{P}|$ being the number of basic blocks $\in \mathcal{P}$.

In contrast, the dynamic invariant path ratio rat_{dyn} computes the WCET contribution of the invariant path to the overall program's WCET:

Definition 4.13 (Dynamic invariant path ratio) The *dynamic invariant path ratio rat_{dyn}* indicates how many WCET cycles were consumed by blocks b of program $\mathcal{P}$

on the invariant path with respect to the overall WCET:

$$rat_{dyn} = \frac{\sum_{b \in \mathcal{P} } WCET_{est}(b) \cdot IP(b)}{\sum_{b \in \mathcal{P}} WCET_{est}(b)}$$

The computed static ratio for the 42 benchmarks ranges between 74.1% and 77.9% for code optimized with optimization levels $O0$ up to $O3$. For the dynamic invariant path ratio, it was observed that between 85.4% and 88.8% of the estimated WCET cycles are spent for the execution of the code declared as invariant path. It is obvious that the dynamic ratio is in general larger than the static ratio since repeatedly executed code, e.g., loops or functions lying on the invariant path, contributes multiple times to the dynamic ratio.

Such results underline the optimization potential of the invariant path paradigm. Since more than 85% of the benchmarks' total WCET is contributed by code classified as invariant path, only a small fraction of the code (contributing 15% of the total WCET) is prone to a WCEP switch. Exploiting this knowledge during WCET minimization allows a reduction of redundant timing analyses to update the worst-case timing data. Hence, compared to other widely used WCET-aware optimizations based on a pessimistic WCEP recomputation, the optimization run time can be significantly reduced.

To show the practical use of WCC's invariant path paradigm, the optimization WCET-aware loop unswitching was developed. Its combination with the novel paradigm demonstrates the benefits concerning the optimization run times. The next section is devoted to WCC's loop unswitching.

4.6.6 Case Study: WCET-Aware Loop Unswitching

The source code compiler optimization *loop unswitching* is a well-known control flow transformation aiming at an ACET reduction. It moves loop-invariant condition branches outside of the loop [Muc97]. In case of an `if-else`-statement, the loop body is replicated and placed inside the respective *then-* and *else*-part. The benefits of the optimization are the reduced number of executed branches improving pipeline behavior and more opportunities for parallelization of the loop [BGS94]. An example for loop unswitching is shown in Fig. 4.28.

However, the loop replication comes at the cost of code expansion. For this reason, loop unswitching can not be applied exhaustively to all potential loop candidates if strict code size constraints, as often found in the embedded system domain, must be met. Thus, similar to loop unrolling (cf. Sect. 4.5), a trade-off between the execution time improvement and the resulting code size increase is required. To do so, potential unswitching candidates should be evaluated before being optimized, enabling to unswitch only those loops that promise a high run time improvement while keeping the code size increase minimal.

Due to the lack of information about execution times and frequencies of the code under analysis, an effective unswitching is, however, not feasible in most compilers.

```
for ( i=0; i<100; i++ ) {                    if (w)
  x[i] = x[i] + y[i];                          for ( i=0; i<100; i++ ) {
  if ( w )                                       x[i] = x[i] + y[i];
    y[i] = y[i] * 2;           →                 y[i] = y[i] * 2; }
  else                                         else
    y[i] = 1;                                    for ( i=0; i<100; i++ ) {
}                                                  x[i] = x[i] + y[i];
                                                   y[i] = 1;
                                                 }
```

Fig. 4.28 Example for loop unswitching

Typically, the optimization traverses the program in a *depth-first search* order and transforms each found loop as long as a predefined maximal code size restriction is not exceeded. Using this strategy, an efficient exploitation of the unswitching potential can not be guaranteed.

Compared to standard unswitching, WCC's WCET-aware loop unswitching differs in two ways. First, the optimization focuses not on an ACET but on a WCET reduction. Second, the sophisticated infrastructure of WCC including detailed worst-case timing semantics enables an effective unswitching. Each loop candidate is virtually unswitched to evaluate the impact of the transformation. Based on this knowledge, loops promising a high WCET reduction are unswitched first while the transformation of loops with little positive impact on the WCET is postponed and possibly omitted.

Moreover, WCC's loop unswitching addresses the problem of switching WCEPs by incorporating the invariant path paradigm. A WCEP switch may occur during the optimization due to code restructuring which has a hardly predictable impact on the sensitive memory systems of modern processors. The reasons for a potential path switch are twofold. On the one hand, I-caches might show a different behavior w.r.t. the incurred cache misses. By shifting the selection statement, the *then-* and *else*-part are mapped to different addresses in the memory compared to the code before the transformation. Consequently, this code is fetched into different cache lines and may result in additional/reduced cache misses during the execution of other mutually exclusive non-WCEPs which turn into the new longest paths.

On the other hand, many processors incur a penalty when performing a fetch to a misaligned target instruction. In this case, the processor stalls since another access to memory is required to complete the instruction fetch. In literature, this phenomenon is known as the *line crossing effect* and may cause a path switch after a code modification [ZKW+05]. In the context of unswitching, shifting code rearranges the memory layout and may introduce additional/reduced misaligned instructions on the non-WCEPs that yield a path switch.

Exploiting the invariant path paradigm, WCC is capable of classifying program loops into those which may entail a WCEP switch after loop unswitching and those where a path switch can be definitely excluded. The optimization of loops from the second class eliminates the need for pessimistic WCEP recomputations, thus significantly accelerates the WCET-aware optimization.

Algorithm 4.5 Algorithm for WCET-aware loop unswitching

Input: $\mathcal{P}, maxFactor$
Output: $optimized\ \mathcal{P}$
 1: $maxSize \leftarrow codeSize(\mathcal{P}) \cdot maxFactor$
 2: $WCETAnalysis(\mathcal{P})$
 3: $set\langle Loop\rangle\ UC \leftarrow unswitchingCandidates(\mathcal{P})$
 4: **while** $UC \neq \emptyset$ **and** $\exists uc \in UC : WCET(uc) > 0$ **do**
 5:     $allOnIP \leftarrow checkInvariance(UC)$
 6:     **repeat**
 7:        $loop_{fittest} \leftarrow fittest(UC)$
 8:        $unswitching(loop_{fittest}, maxSize)$
 9:        $UC \leftarrow UC\ /\ loop_{fittest}$
10:     **until** $UC == \emptyset$ **or** $allOnIP == false$
11:     $WCETAnalysis(\mathcal{P})$
12: **end while**
13: **return** $\mathcal{P}$

WCC's WCET-aware loop unswitching is depicted in Algorithm 4.5. It begins with the calculation of the maximally permitted code size after unswitching (line 1). After a WCET analysis of program $\mathcal{P}$, worst-case timing data is made available at the source code level using WCC's back-annotation (line 2). In the next step, all unswitching candidates available in $\mathcal{P}$ (even $\notin$ WCEP) are collected in UC (line 3). The subsequent outer loop of the algorithm (lines 4–12) is executed as long as there is at least one unswitching candidate lying on the WCEP. From the set UC of candidates, the *fittest* loop promising the highest WCET reduction is selected (line 7) and unswitched if code size constraints are not violated (line 8). The fitness of a loop is determined by the execution counts of the loop-invariant condition branch, with higher frequencies representing fitter loops. If two branches have the same execution count, the one with the larger WCET is preferred. In the rare case of even equal WCETs, the branch with the smaller code size is selected.

The termination condition of the inner loop of Algorithm 4.5 (line 10) checks whether all collected unswitching candidates are on the invariant path (see *allOnIP*). This test is mandatory for an effective WCET minimization since local changes in the CFG may have an impact on timing properties of another portion of the CFG as mentioned previously.

This effect is illustrated in Fig. 4.29 where mutually exclusive parts of the CFG are denoted as *segments*. The CFG represents a situation as could be possibly found at the beginning of unswitching. For the sake of simplicity, the CFG nodes represent either single basic blocks or loops. Gray nodes mark unswitching candidates with their execution counts. The WCEP traverses all nodes except for node $UC3$. WCC' loop unswitching would begin with node $UC1$ which has the highest execution count. After unswitching this node in segment 1, the modified memory layout may affect segment 2 such that a WCEP switch occurs from node G to $UC3$. Thus, in the next step, not the previously collected loop $UC2$ but $UC3$ should be consid-

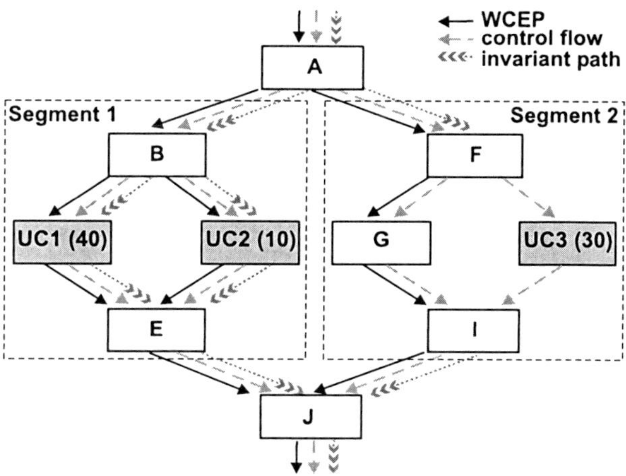

Fig. 4.29 WCEP switch in different segments

ered for unswitching. Obviously, this information is provided after another WCET analysis following the first transformation (line 11).

Within Algorithm 4.5, variable *allOnIP* is responsible for the indication of potential WCEP switches among different CFG segments by forcing a new WCET analysis to keep relevant unswitching candidates updated.

At first glance, checking the invariance of available optimization candidates might seem restrictive, preventing an extensive exploitation of the invariant path. However, as shown in Sect. 4.6.5, a large fraction of the code contributes to the invariant path. Thus, *allOnIP* can be expected to be often *true* such that multiple optimization steps can be performed without a pessimistic WCEP recomputation.

It should be also noted that *allOnIP* is only relevant for code size critical optimizations that aim at a reduction of the WCET while keeping the increase of the code size minimal, and where side effects on the memory system are crucial. Like for WCC's WCET-aware loop unswitching, always the most promising optimization candidate for a WCET minimization should be chosen to maximally improve the program's worst-case performance while keeping the code size increase minimal.

However, for many real processors the non-local effects on the memory system (e.g., line crossing or I-cache misses) are negligible. Moreover, many optimizations focus on a maximal WCET improvement accepting code expansions. For these two common cases, the invariant path information can be fully exploited. Since the optimization of the invariant path always reduces the WCET, the compiler can iteratively optimize these parts of the CFG without performing further WCET analyses. Due to a possible WCEP switch in the remaining code, the WCET information must be once updated for the transformed code to check if possibly new optimization candidates occurred that can be optimized next.

Due to the potentially omitted WCET analyses during the optimization, no reliable assumptions about the WCET of the transformed code can be made. Thus, to get a safe WCET estimation of the optimized program, a final WCET analysis should be performed.

Table 4.3 Benchmark characteristics

Benchmark	#Candidates	Code Size [Byte]	Description
transupp	19	7224	JPEG transformation
wrbmp	6	682	JPEG conversion
block	8	16050	H264 decoding
macroblock	12	18520	H264 decoding

4.6.7 Experimental Results for WCET-Aware Loop Unswitching

To show the effectiveness of the novel paradigm, experiments for WCC's invariant path-based WCET-aware loop unswitching were conducted on real-life benchmarks. The benchmarks stem from the widely used MediaBench suite representing different applications typically found in the embedded systems domain. They contain typical kernel routines that are frequently used in larger benchmarks, e.g., the JPEG-2000 image and the H.264 video compression. The characteristics of the four most interesting kernel routines employed for the evaluation are listed in Table 4.3. The second column of the table indicates the number of candidates for loop unswitching. For the following experiments, an additional code size increase of 50% was allowed, i.e., *maxFactor* = 1.5 (cf. Algorithm 4.5).

4.6.7.1 Optimization Run Time

To demonstrate the positive impact of the invariant path paradigm, the optimization run times of WCET-aware loop unswitching were measured once with and once without exploiting invariant path information. The optimization run time encompasses the entire optimization process from parsing the source code to the generation of the optimized assembly code. Typically, most of the time is consumed by repetitive WCET analyses.

The results depicted in Fig. 4.30 were generated on an Intel Xeon 2.13 GHz system with 4 GB RAM. The 100% mark corresponds to the optimization time of standard loop unswitching without any WCET heuristics. The diagram shows the results for the application of WCC's WCET-aware loop unswitching without exploiting the invariant path information (light bars) and when the invariant path in taken into account (dark bars). It can be seen that for all benchmarks the optimization run time could be drastically reduced when the invariant path information is involved. On average, the run time using invariant path could be reduced by 58%. The conventional WCET optimization without employing the new paradigm took on average 872% more run time than standard WCET-unaware unswitching. Taking the invariant path into account reduces the optimization run time to 379% of standard unswitching which corresponds to a speed-up by a factor of 2.3.

These significant optimization run time improvements result from the reduced number of performed WCET analyses. For example, the number of mandatory

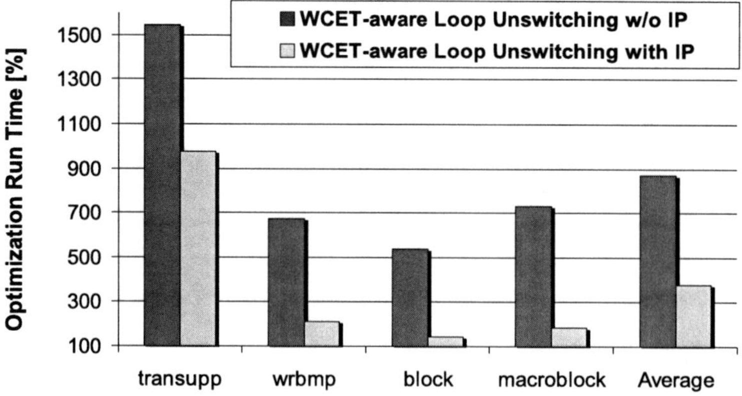

Fig. 4.30 Relative optimization run time without and with invariant path

Fig. 4.31 Relative WCET estimates for WCET-aware loop unswitching

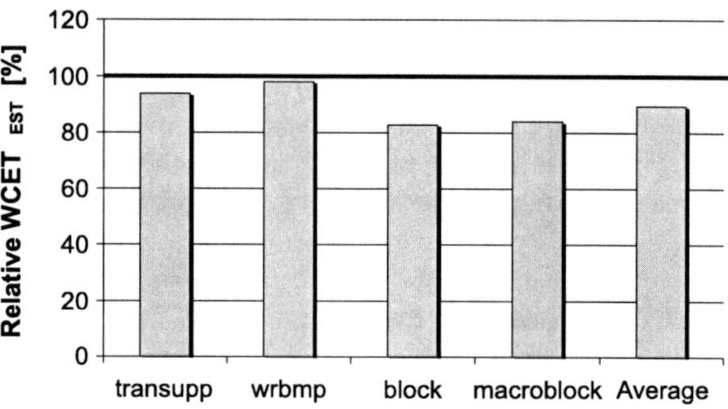

WCET analyses to update the WCEP which was performed for the benchmark *macroblock* could be reduced from seven analyses down to two analyses, reducing the optimization time from more than 75 minutes to less than 19 minutes.

4.6.7.2 WCET

Figure 4.31 presents the WCET reduction achieved by WCC's WCET-aware loop unswitching, with 100% corresponding to the WCET estimation of the original code after dead code elimination. All tests were performed assuming an 8 kByte I-cache of the TriCore processor. Note that the default cache capacity of 16 kByte was reduced to take cache effects for the benchmarks under test into account. As can be seen, an average WCET reduction of 10.4% was achieved for all benchmarks. The maximal WCET reduction of 17.3% was achieved for the *block* kernel. It contains seven loop-invariant `if-then-else`-statements executed between 4 and 16 times in the worst case. By unswitching, their execution frequencies could be significantly reduced.

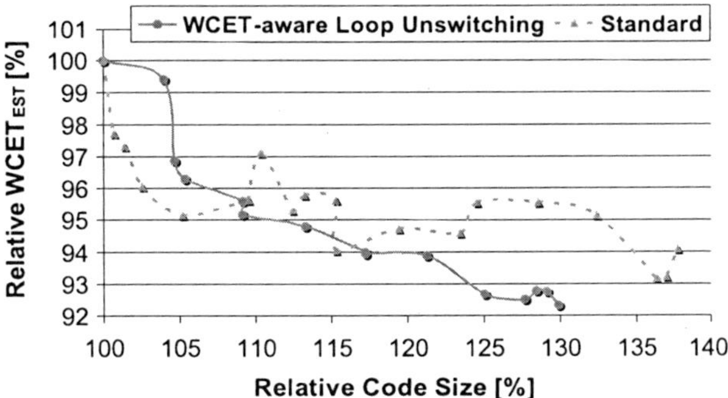

Fig. 4.32 Comparison of standard and WCET-aware unswitching for *transupp*

4.6.7.3 Code Size

The drawback of loop unswitching is the code size increase. For the four benchmarks, an average increase of 19.7% was measured. To bound the code expansion, the WCET-aware loop unswitching may be terminated as soon as the maximally permitted code size increase is achieved.

The diagram in Fig. 4.32 shows the relationship between the WCET reduction and the code size increase for each optimization step of unswitching for the benchmark *transupp*. The solid curve represents the measurements for the WCET-aware loop unswitching, while the dotted curve depicts the measurements for standard unswitching. The points of the curves represent the relative WCET estimation and the relative code size w.r.t. the original code (after dead code elimination) determined after each unswitching of one loop. As can be seen, the WCET-aware optimization transformed 14 out of the 19 unswitching candidates, i.e., the optimization heuristic decided that unswitching of 5 loops does not promise a WCET reduction.

A comparison between the solid curve and the dashed curve reveals that the WCET-aware approach is overall more successful in optimizing the WCET than standard unswitching, since in almost each step a reduction of the WCET estimation was achieved. Only in the last optimization steps a minimal WCET increase of 0.2% was observed. In contrast, standard unswitching randomly finds better loop candidates in the first steps than the heuristic of WCC's WCET-aware unswitching. However, in the following steps, the standard approach results in a noticeable WCET increase after unswitching a selection statement. For example, a relative code size of 105.2% with standard unswitching results in a relative WCET of 95.1%. After unswitching of two further loops, the code size increases to 110.1% with a simultaneous negative effect on the WCET which is increased to 97.1%. This points out that the new optimization is tailored towards an effective WCET reduction and outperforms the standard loop unswitching for this objective.

Moreover, the solid curve can be used by system designers to find a suitable strategy for unswitching which satisfies timing and code size constraints imposed on the system. Based on that curve, the parameters for the desired optimization objective can be extracted and used to terminate the optimization when the desired goals is accomplished.

Ideas from this section have been published in [LGM09].

4.7 Summary

This chapter presents novel compiler optimizations applied at source code level that aim at a reduction of the worst-case execution time. In contrast to traditional compiler optimizations, the proposed techniques exploit detailed timing information of the program under analysis. The WCET data computed by a WCET analyzer at assembly level is made available at source code level using a technique called back-annotation. This timing information enables a quantification of effects of code transformations on the estimated WCET. Hence, an effective improvement of the program's worst-case behavior can be achieved. All the proposed techniques were implemented in the WCC framework and evaluated on real-life benchmarks.

The first optimization studied in this chapter was procedure cloning which may significantly improve the WCET estimation as the optimization allows a more precise specification of flow facts. To cope with the inherent code size increase during cloning, the developed WCET-aware procedure cloning transforms only those functions that promise an improved WCET estimation. An average WCET reduction of 57.5% with a simultaneous code size increase of only 109.5% was achieved on real-life benchmarks.

The second presented optimization is based on superblocks which represent traces consisting of multiple basic blocks. Compiler optimizations benefit from such a code structure since it extends the optimization scope, thus enables new optimization opportunities. In contrast to previous works, the proposed superblocks are constructed at source code level and rely on the worst-case execution path. Combining these WCET-aware superblocks with the compiler optimizations common subexpression elimination and dead code elimination yielded an average reduction of the estimated WCET of 10.2% and 8.8%, respectively, for a total of 55 benchmarks.

WCET-aware loop unrolling was the third proposed optimization. It heavily exploits the back-annotation to get detailed information about the code size, I-cache properties, spill code, and the WCETs. Combining this data with loop iteration counts from a static loop analysis allows the determination of individual unrolling factors for each loop that promise the highest WCET reduction. For a total of 45 benchmarks, an average WCET reduction of 10.2% was observed while standard loop unrolling found in modern compilers led to a WCET reduction of only 2.9%.

Finally, the invariant path paradigm was proposed to accelerate WCET-aware compilation. Optimizations focusing on a WCET minimization must be aware of WCEP switches during code transformations. Typically, the timing information is kept up-to-date by pessimistically recomputing the WCEP after each code modification. However, a WCEP switch can not occur in many situations. Using the invariant path, such situations are detected and redundant WCET analyses can be avoided. To demonstrate the practical use of this new concept, it was integrated into the proposed WCET-aware loop unswitching. Using the invariant path led to a reduction of the optimization run time by 58% on average for the considered benchmarks. It was also shown that the invariant path paradigm can be applied to a large fraction of typical embedded software, making it best suited for a large number of WCET-aware optimizations.

All the techniques presented in this chapter are based on standard ACET optimizations available in many optimizing compilers. Using the proposed techniques, the traditional optimizations can be adapted to minimize the WCET effectively. As the comparison between the standard and the WCET-aware optimizations clearly shows, significantly higher WCET reductions can be achieved when the novel techniques are applied. The reason is the systematic optimization of the crucial parts of the WCEP that promise the largest WCET reductions. As a conclusion, it can be stated that it is worthwhile to study well-known compiler optimizations at source code level, looking for opportunities to trim them for an explicit improvement of the program's worst-case behavior.

Chapter 5
WCET-Aware Assembly Level Optimizations

Contents

5.1 Introduction

In the previous chapter, the eminent role of compilers for code generation of high-performance embedded systems was highlighted. Compiler optimizations can be applied at different abstraction levels of the code, and each level has its inherent strengths. The optimization techniques discussed so far operate at source code level and benefit from the following characteristics: portability, early application in the optimization sequence to enable subsequent optimizations, and availability of more details about the program structure due to the high level of abstraction.

The major shortcoming of source code optimizations is their lack of intrinsic knowledge about the underlying architecture. Hence, the development of transformations that exploit processor-specific features is limited or even infeasible at all. As a result, a maximal optimization potential can not be explored. In contrast, assembly level optimizations operate on a code representation that reflects the finally

P. Lokuciejewski, P. Marwedel, *Worst-Case Execution Time Aware*   131
Compilation Techniques for Real-Time Systems, Embedded Systems,
DOI 10.1007/978-90-481-9929-7_5, © Springer Science+Business Media B.V. 2011

executed code. Thus, the compiler is fully aware of numerous critical details about the utilized resources during execution.

For example, the assembly code specifies which instructions are executed in which order. A compiler can exploit this information to reorder instructions such that the optimized instruction sequences efficiently utilize all processor functional units during execution. Furthermore, thorough knowledge about the executed instructions allows an effective exploitation of the memory hierarchy. Program code and data can be precisely partitioned and systematically allocated to different on-chip memories, such as scratchpads or caches.

The full control of assembly level optimizations on the program code has another advantage. Source code optimizations often rely on fuzzy assumptions about the final machine code since they have little or even no impact on the code selection and succeeding assembly level optimization. Thus, there is no guarantee that assumptions, for instance, about the final memory layout or the utilization of registers are valid, possibly yielding misleading optimization decisions. In contrast, assembly level optimizations operate on code after code selection for which the impact of a modification can be estimated more accurately. Therefore, it can be concluded that any effective optimizing compiler should improve program code at both source code and assembly level to achieve a high code quality for embedded systems.

In this chapter, WCET-aware assembly level optimizations performed in WCC's compiler backend LLIR are discussed that were developed in the course of this book. In Sect. 5.2, a survey of existing assembly level optimizations for hard real-time systems is provided. Section 5.3 introduces the optimization procedure positioning which improves instruction cache (I-cache) behavior along the WCEP by a suitable re-arrangement of procedures in memory. To expose a high instruction-level parallelism in the code, Sect. 5.4 presents a global instruction scheduling technique called trace scheduling that reorders instructions across multiple basic blocks. Finally, this chapter is concluded in Sect. 5.5.

5.2 Existing Code Optimization Techniques

Assembly level is the primary domain for WCET-aware compiler optimizations. Typically, a WCET analysis is performed at this level, enabling an easy access to WCET data. An advanced compiler infrastructure, such as WCC's back-annotation (cf. Sect. 3.7), that translates timing information into other code representations is not required.

The major class of WCET-aware assembly level optimizations aims at an improved utilization of the processor's memory hierarchy system. Several works focus on caches. Software-controlled caches enable locking of cache lines to avoid evictions. Campoy et al. [CPI+05] used a genetic algorithm for static I-cache locking, which may, however, not generate optimal results. Cache locking based on an explicit consideration of the WCEP was presented in [FPT07]. The authors performed an iterative approach that decides in each optimization step which parts of the code should be locked in the I-cache to achieve a high WCET reduction compared to the

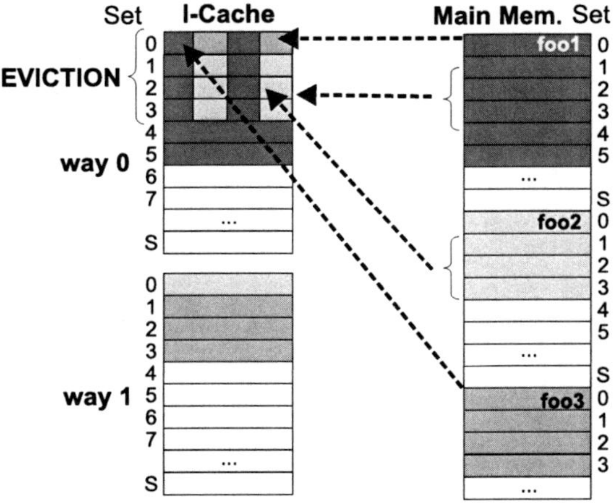

Fig. 5.2 Example for cache content eviction before positioning

into the cache, beginning at way 0 of cache set 0. Subsequently, function `foo2` is invoked. Since, way 0 of set 0 is already occupied, the first block of `foo2` is copied into the free way 1 of set 0, while the remaining code is copied into ways 0. The execution of `foo3` leads to the first eviction in set 0 since both ways are already occupied. Due to the LRU replacement strategy, way 0 is chosen. The remaining code is moved into the free blocks of way 1 of sets 1–3. Finally, the remaining code of `foo1` is executed. Due to missing free ways in both sets, it evicts blocks from way 0 in sets 1–3, copying the remaining two memory blocks into the free blocks of cache way 0 of sets 4–5. For the remaining loop iterations of *foo1*, the eviction of cache ways is continued. This results in multiple *cache conflict misses* which entail multiple accesses to the slow external memory.

The costly eviction of cache blocks can be eliminated by altering the order in which functions are located in memory. In general, this is accomplished by allocating functions which are accessed within a local time window (temporal locality) contiguously in memory as depicted in Fig. 5.3. Function `foo2` is mapped at a memory address corresponding to set 6. Obviously, locating `foo3` contiguously to `foo1` would have the same result. This eases the pressure of mapping multiple memory locations to the same sets. Executing the code from Fig. 5.1, the entire code for the three functions can be brought into the cache. This memory layout eliminates all set evictions, thus allowing a fast execution due to cache content reuse.

For *direct-mapped* caches, the positioning technique might be even more beneficial. In this cache architecture, each set can be considered as holding exactly one way. Hill and Smith [HS89] reported that direct-mapped I-caches may show a larger number of conflict misses compared to set-associative caches. Hence, altering the order of the code in the described manner would eliminate potentially more conflict misses.

It should be noted that the term *function* was used in this example to be consistent with ANSI C terminology. Since WCC's procedure positioning algorithms

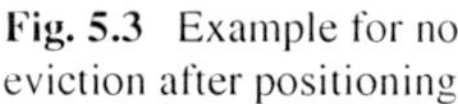
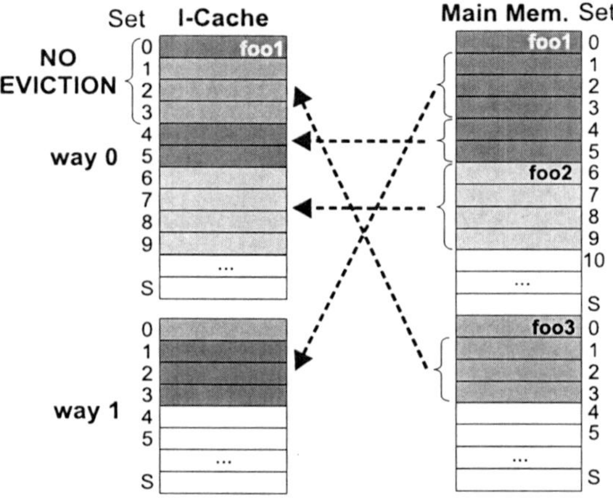

Fig. 5.3 Example for no eviction after positioning

are performed at assembly level, the term *procedure* will be used as a synonym for *function* in the following.

5.3.2 Related Work

A large number of ACET optimizations exploit memory hierarchies and aim at the improvement of both data and instruction cache behavior. The main idea behind these techniques is to enhance spatial and temporal locality. Popular optimizations for data caches encompass *loop interchange, loop tiling, loop fusion* or *data prefetching* [Muc97].

I-caches mainly benefit from a reorganization of the code at procedure and basic block level. Tomiyama [TY97] proposed two code placement methods for basic blocks to reduce the cache miss rate based using ILP. Hwu and Chang [HC89] proposed a compiler with an integrated instruction placement algorithm reducing page faults. The compiler first groups basic blocks which tend to be executed in sequence into traces, maximizing the sequential and spatial locality. In a second step, procedures which are executed close to each other in time are placed into the same page. Thus, inter-function intersections are reduced which also remove potential cache conflicts among interacting functions.

In [LW94], a cache profiling system identifies hot spots by providing cache performance information at source code level. After an automatic classification into compulsory, capacity and conflict misses, the profiler suggests appropriate standard program transformations to improve cache performance. The approach presented by Mendlson et al. [MPS94] does not rely on profiling data but on static information. In addition, unlike previously cited works, their approach requires the exact knowledge of the cache architecture. Their idea is to prevent different segments of code executed in a loop to be mapped into the same cache area by code replication.

Static cache analysis is essential for a WCET analysis of cache-based processors. Its goal is to classify each memory access as a cache hit or a miss. Ferdinand used *must* and *may analysis* based on abstract interpretation [FHL+01]. The former determines if a memory access is always a cache hit while the latter computes if the access may be a hit. This approach is also used in aiT, the WCET analyzer integrated into WCC.

The only work that exploited procedure positioning for WCET reduction was discussed in [LJC+10]. The authors adopted techniques presented in this section to optimize applications running on a communication gateway of a wireless body-area sensor network (BAN). By deriving a good order of different sensor procedures, not only the WCET but also the power consumption could be reduced.

A related technique to procedure positioning is code positioning of basic blocks and was employed by Zhao et al. [ZWH+05] to reduce the WCET. They exploited WCET data to re-arrange the memory layout of basic blocks in such a way that branch penalties along the WCEP were avoided. The main difference between Zhao's work and WCC's WCET-aware procedure positioning is that Zhao focused on the reduction of incurred pipeline delays caused by control transfer instructions, while the techniques discussed in this section eliminate cache conflict misses.

5.3.3 *Standard Procedure Positioning*

Standard procedure positioning operates on a call graph:

Definition 5.1 (Call graph) A *call graph* is a weighted, undirected graph $G = (V, E)$, where nodes V correspond to procedures and edges $E \subseteq V \times V$ connect two nodes $v_i, v_j \in V$, iff v_j is invoked by v_i. An edge weight $w(e_{ij})$, with $e_{ij} = (v_i, v_j) \in E$, represents the call frequency, i.e., how often v_i and v_j invoke each other mutually during program execution.

Typically, procedure positioning aims at ACET reduction. To obtain call frequencies for the call graph, profiling is performed. Based on the call graph, procedures that are combined by an edge with a high weight should be located contiguously in memory. This way, mutual cache eviction of procedures with a high call frequency can be reduced.

Procedure positioning has several positive effects on the system. Besides the already discussed benefits concerning the performance of the I-cache, the altered memory layout after positioning may also eliminate *translation lookaside buffer (TLB) misses* [PH90] that are often used in general purpose systems. Reorganizing ⁃⁃⁃tions contiguously in memory increases the probability that both functions will ⁃he same page. Hence, the page working set and potentially TLB

another advantage which is especially relevant for em-
ant objective for embedded systems is energy efficiency
en powered by a battery. Code positioning increases the
moving cache set eviction, thus eliminating cache misses.

In contrast to a cache hit, which corresponds to a cache read, the energy model for
a cache miss is given as follows [VM07]:

$$E_{Cache\_miss} = 2 * E_{read}(Cache)$$
$$+ \, linesize(Cache) * (E_{read}(MM) + E_{write}(Cache))$$

where $E_{read}(Cache)$ and $E_{write}(Cache)$ are the energies consumed by the cache
memory for a read and a write access, respectively. Moreover, $E_{read}(MM)$ represents the energy consumption for a read access to main memory. In case of a cache
miss, two read accesses are required: the first read identifies the miss while the second read finally sends data to the processor once it has been brought into the cache.

A decreased number of cache misses results in less accesses to the main memory.
Verma [VM07] reported that the energy consumption for an access to main memory
compared to accessing a cache may be increased by a factor of 40. Hence, code
positioning may produce code that substantially saves energy consumption.

However, the traditional ACET positioning may be ineffective for WCET reduction. If the edge weights of the call graph gathered during profiling differ from the
worst-case call frequencies computed during WCET analysis, then inappropriate
optimization steps may be conducted. Due to this reason, a WCET-aware procedure
positioning was developed that explicitly relies on WCET information.

5.3.4 WCET-Centric Call Graph-Based Positioning

In contrast to standard, profiling-based optimizations, WCC's novel positioning extracts input data for the call graph from the WCET analyzer aiT. This fundamental
difference makes WCC's approach not prone to changing inputs. Profiling data is
critical since it reflects the program execution for a particular set of input data, i.e.,
profiling the program under test with varying inputs may yield different results. For
more complex programs that consist of numerous input-dependent execution paths,
it is almost infeasible to find representative input values. This may lead to a call
graph that is annotated with profiling data that does not cover particular program
executions. The optimized code will possibly not improve cache behavior and may
even suffer from a performance degradation.

WCC's positioning does not rely on representative input data. Edge weights are
computed by a WCET analyzer and are invariant for all program executions. These
frequencies are used for the construction of WCC's WCET-centric call graph:

Definition 5.2 (WCET-centric call graph) A *WCET-centric call graph* is a call
graph $G = (V, E)$ where edge weights correspond to worst-case call frequencies,
i.e., the number of mutual invocations between two procedures on the WCEP.

Those edges with the heaviest weight potentially combine the most prom
functions for optimization.

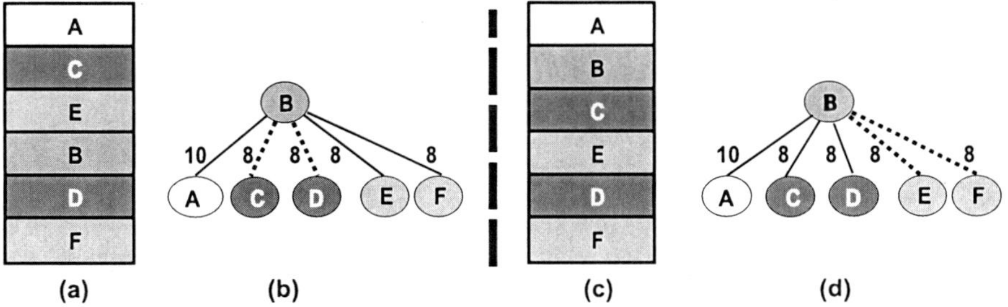

Fig. 5.4 Inappropriate positioning based on profiling

5.3.4.1 Greedy WCET-Aware Positioning Approach

It is well known that the impact of memory layout modifications on caches is hardly predictable. Therefore, a greedy approach that evaluates the impact of a particular procedure rearrangement on the WCET seems promising. In case of an achieved WCET reduction, the new memory layout is considered as a new starting point for the next optimization cycle, and the next most promising function for positioning is considered. Hence, the approach successively reduces the WCET and guarantees that no degradation of the WCET is accepted.

WCC's greedy approach is tailored towards a WCET reduction and the procedure positioning layout might substantially differ from the layout chosen by the existing ACET positioning optimizations. This situation is illustrated in the following.

Example 5.1 The following figure demonstrates why procedure positioning guided by profiling data may be inappropriate for WCET reduction and might even yield a decreased worst-case performance.

Assume the function memory layout given in Fig. 5.4(a) and the corresponding call graph in Fig. 5.4(b), where the edge weights represent call frequencies obtained during profiling. In addition, the dotted edges indicate the current WCEP. This call graph represents a typical `if-then-else`-statement:

```
void B(...) {
    if (...) {
        A();
    } else
    if (...) {
        C(); D();
    } else {
        E(); F();
    }
}
```

Before optimization, the WCEP traverses the first **else**-part, i.e., procedures **C** and **D** contribute to the WCET. Hence, the WCEP does not correspond to the most frequently executed path which traverses the **if**-part. Based on profiling, ACET code

---

Algorithm 5.1 Algorithm for greedy WCET-aware procedure positioning

---

Input: $\mathcal{P}$
Output: *optimized* $\mathcal{P}$
 1: $WCET_{\mathcal{P}} \leftarrow WCETAnalysis(\mathcal{P})$
 2: $G_{WCET} \leftarrow BuildCallGraph(\mathcal{P})$
 3: $G_{ref} \leftarrow BuildCallGraph(\mathcal{P})$
 4: **repeat**
 5:   **repeat**
 6:     $e_{max} \leftarrow FindMaxEdge(G_{WCET})$
 7:     **if** $e_{max} == \emptyset$ **then**
 8:       **return** $\mathcal{P}$
 9:     **end if**
10:     $possibleWCEPSwitch \leftarrow$ **true**
11:     $Positioning(e_{max}, G_{ref}, \mathcal{P})$
12:     $WCET_{new} \leftarrow WCETAnalysis(\mathcal{P})$
13:     **if** $WCET_{new} > WCET_{\mathcal{P}}$ **then**
14:       $RollbackPositioning(\mathcal{P})$
15:       $G_{WCET} \leftarrow G_{WCET} \setminus e_{max}$
16:       $possibleWCEPSwitch \leftarrow$ **false**
17:     **else**
18:       $WCET_{\mathcal{P}} \leftarrow WCET_{new}$
19:     **end if**
20:   **until** $possibleWCEPSwitch ==$ **true**
21:   $UpdateCallGraph(G_{WCET}, \mathcal{P})$
22: **until false**

---

positioning would allocate functions **A** and **B** contiguously in memory to avoid conflict misses (cf. Fig. 5.4(c)). This might result in new conflict misses on path $\mathbf{B} \rightarrow \{\mathbf{E, F}\}$, outweighing the execution time of the current WCEP $\mathbf{B} \rightarrow \{\mathbf{C, D}\}$. Therefore, a modification of the memory layout may result in a different path becoming the WCEP whose WCET is even larger than that of the original WCEP before the transformation. This leads to a WCEP switch (cf. Fig. 5.4(d)) and an increased overall program's WCET. Thus, a decreased ACET was achieved at the cost of an increased WCET estimation.

To avoid such adverse effects, WCC's greedy approach is an iterative algorithm that processes a single edge of the WCET-centric call graph during each iteration cycle. Each modification of the memory layout is virtually evaluated w.r.t. its impact on the program's WCET, and only beneficial positioning steps are accepted.

WCC's WCET-aware procedure positioning adopts ideas of the profiling-guided positioning described by Pettis and Hansen [PH90]. The optimization is depicted in Algorithm 5.1. The algorithm performs a WCET analysis of program $\mathcal{P}$ and constructs two WCET-centric call graphs based on the worst-case call frequencies (lines

1–3). As will be shown later, the reference call graph G_{ref} remains unmodified during the entire optimization and may be consulted during positioning.

WCET-aware positioning begins with the call graph edge that is weighted with the maximal worst-case call frequency (line 6). If no such edge exists, the algorithm is terminated (lines 7–9). Otherwise, this edge is utilized for positioning of program $\mathcal{P}$ (line 11). Within *Positioning*, procedures specified by e_{max} are reordered in such a way that they are allocated contiguously in memory. In the beginning of the algorithm, all nodes in the call graph G_{WCET} represent single procedures. After choosing two procedures to be placed contiguously in memory, both nodes of G_{WCET} are merged and their edges are coalesced. This situation is illustrated by the following example.

Example 5.2 A possible WCET-centric call graph is depicted on the left-hand side of the following figure:

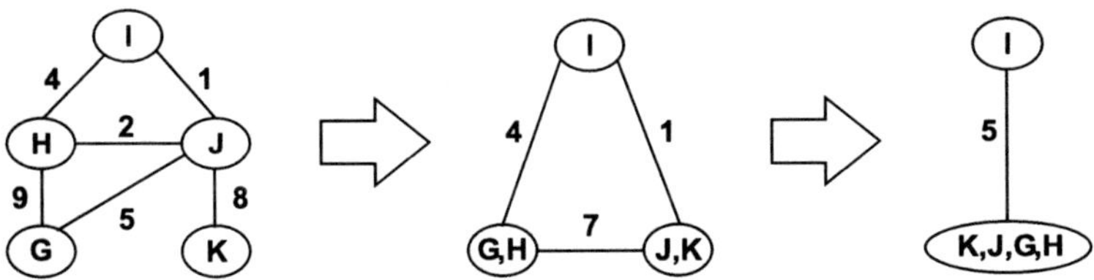

This graph corresponds to G_{ref} in Algorithm 5.1. WCC's WCET-aware positioning would begin the positioning of procedures **G** and **H** in the first step, and procedures **J** and **K** in the next step. If these two memory layout modifications yield an improvement of the estimated WCET, the WCET-centric call graph would be updated as indicated in the middle of the above figure. As can be seen, nodes **G** and **H** as well as **J** and **K** are coalesced. The coalesced nodes guarantee that the contiguous allocation of the corresponding procedures is preserved for further invocations of function *Procedure* in Algorithm 5.1. Moreover, edges are merged by accumulating the corresponding weights.

The decision for an appropriate procedure order becomes more complicated if two already coalesced nodes have to be merged. The second node can be placed before or after the first node. Moreover, the procedure chains (order of procedures within a node) can be reversed within the coalesced nodes, leading to four potential decisions in total. To find the most promising memory allocation, the reference call graph G_{ref} is consulted. It helps to find these two procedures at the end of each chain that exhibit the greatest worst-case call frequency. In case of the example above, the maximal call frequency of 5 is observed between procedures **G** and **J**. Hence, the procedure chain of the second node $\langle$**J**, **K**$\rangle$ is reversed and placed before node $\langle$**G**, **H**$\rangle$.

After positioning, Algorithm 5.1 performs a timing analysis (line 12) and compares the estimated WCET with the WCET before positioning. If positioning had a negative impact on the program WCET, the memory layout modification is rolled back and the current edge is removed from the call graph (lines 13–16). In this case, the algorithm continues with the next edge having the maximal call frequency. Otherwise, if positioning yielded a WCET reduction, the new memory layout is preserved and the call graph is updated (line 21). This update warrants that possible WCEP switches after positioning are taken into account. WCC's positioning terminates if no more promising edges are found (lines 7–9).

5.3.4.2 Heuristic WCET-Aware Positioning Approach

Due to the possibly large number of time-consuming WCET analyses of the greedy approach, also a fast heuristic was developed that just uses the data of the WCET-centric call graph for the initial input program. In contrast to the greedy approach, the heuristic performs exactly one WCET analysis to construct the initial call graph.

Hence, this heuristic approach can be considered as a simplification of the greedy Algorithm 5.1. Instead of evaluating each positioning step w.r.t. the WCET (lines 12–19), the heuristic performs positioning for each edge in the order of the weights.

The speed advantages come at the cost of efficacy. First, the reordering of procedures is based exclusively on the initial call graph and is performed without a re-evaluation of its impact on the WCET. Hence, also undesired WCET increases are accepted. Second, WCEP switches are not considered. Since the call graph is not updated, the heuristic approach operates on an outdated WCET-centric call graph if the WCEP changes. The applied positionings would then possibly not affect the WCET.

5.3.5 Experimental Results for WCET-Aware Procedure Positioning

This section evaluates the impact of WCC's WCET-aware procedure positioning on the WCET estimates of real-life benchmarks. The experiments were conducted for the TriCore TC1796 processor with enabled I-cache. The following cache parameters were used: capacity of 8 kByte, 2-way set associativity with a cache line size of 256 bits, and an LRU replacement strategy. The benchmarks stem from the test suites MRTC [MWRG10] and MediaBench [LPMS97]. Their characteristics are depicted in Table 5.1. The second column indicates the benchmarks' code sizes while the third column shows how many procedures each benchmark contains that are candidates for procedure positioning. These code sizes do not change during optimization since the applied re-allocations of procedures do not require any additional code.

5.3.5.1 WCET

Figure 5.5 shows the results for greedy and heuristic procedure positioning. 100% corresponds to the WCET estimates of the original code of the considered real-life benchmarks compiled with *O3*. It can be seen that a WCET reduction was achieved

Table 5.1 Characteristics of evaluated benchmarks

Benchmark	Size [kByte]	# Functions
expint	0.97	3
g721_encode	19.9	26
g723_encode	19.9	26
gsm_decode	34.1	35
gsm_encode	41.9	36
mpeg2	39.8	14

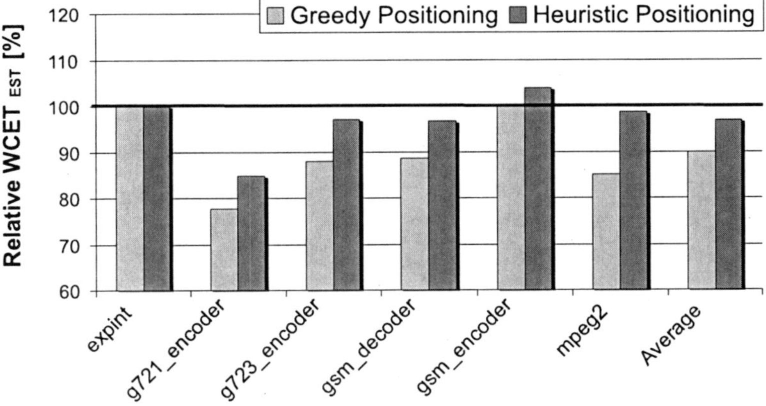

Fig. 5.5 Relative WCET estimates for WCET-aware positioning

for most benchmarks. The greedy algorithm achieved an average WCET reduction of 10.1%, while the heuristic approach reduced the WCET by 3.1% on average.

The results strongly depend on the initial order of the benchmarks' procedures. If the original memory layout already yields a good cache performance, positioning might lead to smaller improvements than for benchmarks that lead to more cache conflict misses. Moreover, tiny benchmarks whose text section is small enough to fit entirely into the cache (e.g., *expint*) do not profit from this optimization since no conflict misses can occur. However, applications that completely fit into the (usually) small I-cache of a resource-restricted embedded system are very uncommon.

For the *gsm_encoder* benchmark, the greedy algorithm could not achieve an improvement and the heuristic approach even slightly worsened the WCET. This is due to the influence of memory layout modifications on the memory system performance which is hardly predictable. A reordering of procedures might improve the cache behavior locally but might simultaneously induce new cache misses for the execution of other code fragments, leading to a degraded overall cache performance. Moreover, modifications to the memory layout may introduce additional line crossing effects (cf. Sect. 4.6.6) that adversely affect the program performance.

For all benchmarks, greedy positioning achieved better results since it does not allow a degradation of the WCET. This might result in a local optimum missing the

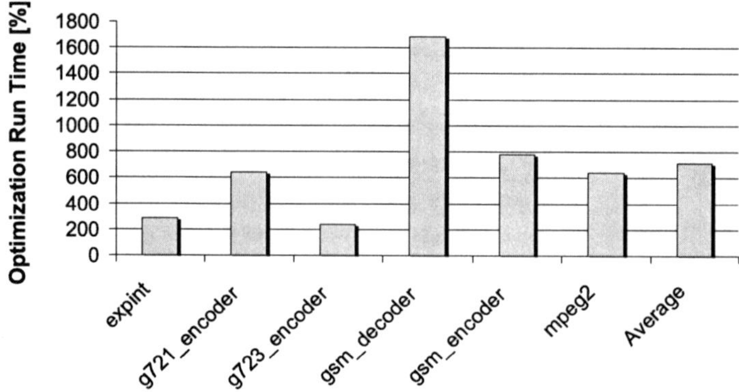

Fig. 5.6 Relative optimization run time for WCET-aware positioning

global minimum as could be potentially achieved by the heuristic approach. However, for the considered benchmarks, this case did not arise. The heuristic approach might worsen the WCET as experienced for *gsm_encoder*. Hence, it is worthwhile to invest time for the optimization to achieve best results, as done by WCC's greedy approach.

5.3.5.2 ACET

The impact of WCET-aware positioning was also measured w.r.t. the ACET of the benchmarks. On average, the greedy approach achieved an ACET reduction of approximately 4%, while the heuristic approach decreased the ACET by less than 2%. These results show that WCC's positioning is tailored towards a WCET reduction which may be less successful for ACET minimization since the call frequency edge weights in the call graph may differ for the average- and worst-case.

5.3.5.3 Optimization Run Time

The optimization run time for WCC's WCET-aware procedure positioning was measured on an Intel Xeon 2.13 GHz system with 4 GB RAM. Obviously, the heuristic approach is faster since it relies on a single WCET analysis. In contrast, the optimization run time of the greedy approach strongly depends on the number of evaluations of memory layout modifications that are accompanied by a WCET analysis. Using the greedy approach, most time was spent for the optimization of *mpeg2* which amounts to 182 minutes. For other benchmarks, the optimization time ranged between 1 and 42 minutes. Hence, the run times are moderate for the optimization of embedded systems.

Figure 5.6 depicts the relative optimization time for the greedy approach, with 100% being the run time for the heuristic approach. As can be seen, the greedy approach may increase the optimization time by up to 1,720% which translates to an absolute optimization time of 25.3 minutes for the benchmark *gsm_decoder*. However, as was seen in Fig. 5.5, this additional overhead pays off.

The techniques presented in this section led to the publication [LFM08].

5.4 Trace Scheduling

The execution order of instructions can have a significant effect on the program execution time. The reasons are manifold. Instructions may take several execution cycles to deliver their result, thus they block other instructions that are data-dependent on them. Accesses to memory may delay the execution of instructions. Moreover, particular combinations of instructions *stall* a processor pipeline, while others allow a simultaneous execution on a superscalar pipeline. Instruction scheduling, also known as code compaction, is a popular assembly level optimization helping to find good instruction orders that enhance the number of instructions executed in parallel.

Local instruction scheduling confines its attention to single basic blocks. However, this limited approach often fails to exploit full instruction-level parallelism since the number of instructions in a block to be scheduled is limited. To circumvent this problem, instruction scheduling can be extended to operate on multiple basic blocks. The latter is called *global scheduling*. A prominent example of a global scheduler is *trace scheduling* which is a profiling-driven method aiming at an ACET reduction. In this section, trace scheduling is studied for an improved WCET.

The remainder of this section is organized as follows. Section 5.4.1 presents a motivating example to indicate the benefits of scheduling across basic block boundaries. An overview of related work on local and global scheduling is provided in Sect. 5.4.2. The approach of local basic block scheduling is discussed in Sect. 5.4.3, followed by a presentation of extensions towards a WCET-aware trace scheduling in Sect. 5.4.4. Finally, experimental results achieved on real-life benchmarks are provided in Sect. 5.4.5.

5.4.1 Motivating Example

Local instruction scheduling is restricted by basic block boundaries. Thus, even if the basic block scheduler finds an optimal schedule, the overall instruction-level parallelism may not be sufficiently exposed. This shortcoming may become a serious concern for superscalar *in-order* issue processors [HP03] that issue multiple instructions in program order.

WCC's target architecture, the Infineon TriCore TC1796, belongs to this class of processors. Its superscalarity is realized by 3 different pipelines that can execute instructions simultaneously: the integer pipeline mainly executes instructions on data registers, the load/store pipeline handles load/store instructions as well as address

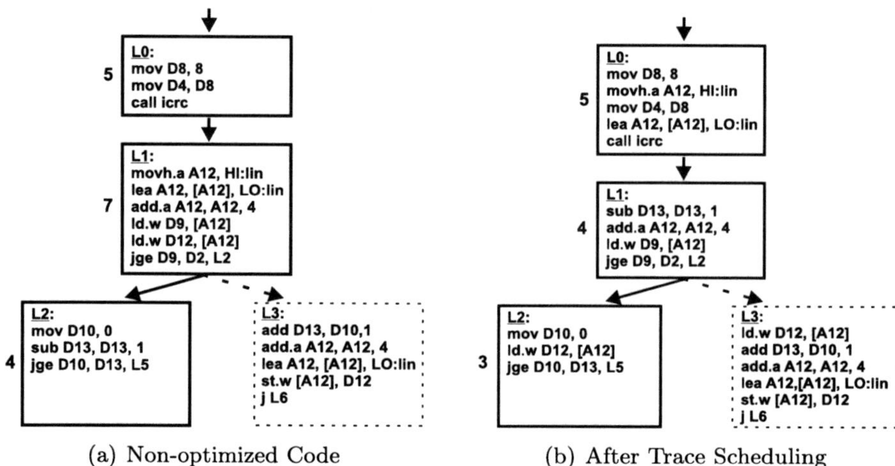

(a) Non-optimized Code   (b) After Trace Scheduling

Fig. 5.7 Example for trace scheduling

computations, and the loop pipeline is responsible for the zero-overhead loop instruction. To exploit this parallelism, the instructions within the program code have to be placed in the code in a special order.

To demonstrate the strength of trace scheduling, consider Fig. 5.7(a) depicting a fragment of the CFG taken from the non-optimized assembly code of benchmark *crc* [MWRG10] using the TriCore instruction set. The solid boxes and arrows represent the WCEP, while block L3 does not contribute to the WCET. Using this instruction order, TriCore's pipelines can not be utilized in parallel since there is no mix of instructions in any block that could be simultaneously issued on the integer and load/store pipeline. Even the local instruction scheduler is not able to generate code that exhibits an increased parallelism. Based on this instruction order, the length of the WCEP amounts to 16 cycles as shown by the numbers next to the basic blocks (for the branch and call instructions, an execution of 2 and 3 cycles, respectively, was assumed).

Unlike local scheduling, trace scheduling allows the motion of instructions across block boundaries. Focusing on the WCEP, instructions are moved across blocks L0, L1, and L2. This way, bundles of instructions are generated that are issued in the same cycle on different pipelines. For example, the two first instructions `mov D8, 8` and `movh.a A12, HI:lin` can be simultaneously passed to the integer and load/store pipeline, respectively. It can also be seen that the motion of the load instruction `ld.w D12, [A12]` below the first conditional jump `jge D9, D2, L2` makes a duplication of this instruction into the non-WCEP block L3 mandatory. Even though the execution time of L3 is increased, the overall length of the WCEP could be shortened from 16 to 12 cycles. If this fragment of the CFG represents a frequently executed kernel code, the reduction of 4 cycles may accumulate to a significant total saving of execution cycles. Hence, this example shows that global scheduling along the WCEP even at the cost of other paths, which are not part of the WCEP, is a promising WCET optimization.

5.4.2 Related Work

Processors without an automatic pipeline interlock control, such as the Infineon PP32 [NGE+99], require an instruction scheduler which adds NOP instructions into the code to ensure correct program execution. However, the primary goal of instruction scheduling is typically the enhancement of the instruction-level parallelism to speed up program execution. Scheduling can be either carried out by hardware at run time [HP03] or by software at compile time.

Compile-time instruction scheduling is crucial for *very long instruction word* (*VLIW*) machines to achieve an instruction-level parallelism of up to 14 operations that these machines are capable of issuing in a cycle [CNO+87]. But also for superscalar in-order issue processors, instruction scheduling is mandatory for high performance. In this book, compile-time instruction scheduling for the latter class of processors is considered.

If instruction scheduling is performed before register allocation, it is referred to as *prepass scheduling*, otherwise *postpass scheduling* [Muc97]. Running scheduling after register allocation may limit the optimization since the program code suffers from *artificial* data dependencies introduced by the register allocator. On the other hand, this approach has the advantage that spill code can be properly scheduled. There are also approaches where instruction scheduling is integrated into the register allocator [NP93, MPSR95]. However, since both problems are complex enough by themselves, the combined approach is typically inapplicable for larger programs. Hence, most compilers treat these transformations separately—this is also done in WCC.

Earlier works concentrated on local instruction scheduling that operates on single basic blocks. However, it was recognized that basic block boundaries are a serious obstacle for achieving good performance. Therefore, global scheduling approaches were developed. These approaches have in common that they attempt to reduce the execution time of the most frequently executed path. A popular example for a global scheduling approach, which works on a set of basic blocks with acyclic control flow (aka. global acyclic scheduling), is *trace scheduling* [Fis81]. Further examples for acyclic approaches are *superblock scheduling* [HMC+93] (cf. Sect. 4.4) as well as *hyperblock scheduling* [MLC+92]. The latter relies on conditional (predicated) instructions. To expose instruction-level parallelism across different loop iterations, cyclic scheduling was developed. Standard techniques are *software pipelining* [RTG+07] and *modulo scheduling* [WHSB92] that overlap the execution of instructions from multiple iterations of the loop.

Instruction scheduling has also been combined with techniques from machine learning. Cavazos and Moss [CM04] exploited supervised learning in a Java just-in-time (JIT) compiler to reduce compilation time by applying local instruction scheduling only to those basic blocks that promise a significant improvement. Moreover, Russell et al. [RMC+09] used supervised learning to automatically generate heuristics for superblock scheduling.

5.4.3 Local Instruction Scheduling

In this section, fundamental ideas behind local scheduling are discussed since these concepts serve as basis for WCC's WCET-aware trace scheduler. Scheduling can be carried out before or after register allocation. For WCC's trace scheduling, executable code is required since this approach relies on worst-case execution counts of CFG edges that enable the construction of a critical path. Hence, the following discussion is based on the assumption that the code under analysis has already been processed by a register allocator.

For processors without an automatic pipeline interlock, instruction scheduling is a mandatory compiler technique to preserve program semantics. In contrast, for processors with automatic locking mechanisms, instruction scheduling serves as an optimization technique, providing the following positive effects on program performance [Muc97, CT04]:

- Avoidance of *pipeline stalls* by hiding memory and functional unit latencies.
- Filling of branch delay slots with useful instructions to hide delays due to control flow modifications [HP03].
- Exploitation of superscalarity by increasing number of simultaneously issued instructions.

Regarding the TriCore architecture, the first two issues are irrelevant for a scheduler. Whenever memory latencies arise during execution, TriCore's pipelines are completely stalled, thus there is no opportunity to hide these latencies. Moreover, most instructions have an execution latency of a single cycle [Inf08a] which keeps the number of functional unit latencies small. Also, this processor is not equipped with branch delay slots that could be filled with useful instructions.

In contrast, the exploitation of superscalarity promises a significant enhancement of instruction-level parallelism. As mentioned in Sect. 5.4.1, the TriCore processor is equipped with three 4-stage pipelines that are dedicated to the execution of particular instruction types. If the instructions are ordered in the program code in bundles, a parallel execution can be achieved. Bundles refer to instruction pairs that consist of one instruction that is issued on the integer pipeline while the other is issued on the load/store pipeline. In addition, the loop pipeline can be utilized for the execution of TriCore's `loop` instruction [Inf08b], leading to a further increase of parallel execution. As will be described in the next section, this strategy is also pursued in WCC's instruction scheduler that aims at maximizing the number of bundles in program code.

Figure 5.8 provides an example for parallel execution of the code depicted on the left-hand side of the figure. The situation in the four stages of the integer and load/store pipeline depending on the execution cycles is depicted on the right-hand side of the figure. While the first instruction `sub D4, D5, D6` is executed separately due to missing bundle candidates, the following instructions ($\langle i2, 11 \rangle, \langle i3, 12 \rangle$) are executed simultaneously. For more details on the behavior of TriCore pipelines, the interested reader is referred to [Inf04].

```
sub  D4, D5, D6        ;i1
add  D7, D8, D9        ;i2
ld.w D10, [A2]         ;l1
add  D11, D12, D13     ;i3
ld.w D14, [A3]         ;l2
```

		Execution Cycles					
		1	2	3	4	5	6
Integer Pipeline	Fetch	i1	i2	i3			
	Decode		i1	i2	i3		
	Execute			i1	i2	i3	
	Writeback				i1	i2	i3
Load/Store Pipeline	Fetch		l1	l2			
	Decode			l1	l2		
	Execute				l1	l2	
	Writeback					l1	l2

Fig. 5.8 Parallel execution on TriCore pipelines

5.4.3.1 List Scheduling

To generate a schedule, compilers require a representation of the code that describes relevant dependencies between instructions in order to preserve program semantics. For this purpose, a *dependence graph* $\mathcal{D}$ can be employed:

Definition 5.3 (Dependence graph) A *dependence graph* $\mathcal{D} = (V, E)$ of basic block b is a directed acyclic graph, where nodes V correspond to instructions $i \in b$ and edges $E \subseteq V \times V$ connect two nodes $v_i, v_j \in V$ iff there is a data dependency between v_j and v_i. The dependencies can be of the following type:

- True dependency (RAW): v_i writes an operand that is read by v_j.
- Anti-dependency (WAR): v_j writes an operand that is read by v_i.
- Output dependency (WAW): v_i and v_j write the same operand.

If the last instruction $i_{last} \in b$ is a branch or call instruction, edges between all other instructions and i_{last} are added to preserve control flow. Moreover, each edge $(v_i, v_j) \in E$ is associated with a label that is the execution latency between instructions v_i and v_j.

The dependence graph can be constructed in a top-down manner by beginning with the first instruction of a basic block b. Whenever a new instruction is considered, a corresponding node is added to $\mathcal{D}$. This new instruction is compared against all previous instructions and identified dependencies are modeled as edges between the respective nodes. The comparison is repeated for all instructions in b. In total, this approach requires $O(n^2)$ steps, where n is the number of instructions in b.

Given a dependence graph $\mathcal{D}$ for a basic block, a schedule S maps each node $v \in V$ to a non-negative integer that represents the cycle in which v should be issued. Since optimal local instruction scheduling is $\mathcal{NP}$-complete [PS93], compilers generate approximate solutions to the instruction scheduling problem using greedy heuristics. In practice, most scheduling algorithms are based on a well-known technique called *list scheduling* [MG04]. This approach has been the dominant paradigm

---

Algorithm 5.2 List scheduling

Input: *Dependence Graph $\mathcal{D}$*
Output: *Instruction Schedule S*
 1: *list < Instructions > ReadyList ← DetermineSourceNodes($\mathcal{D}$)*
 2: *cycle ← 0*
 3: **while** *ReadyList ≠ ∅* **do**
 4:     *AssignPriority(ReadyList)*
 5:     *SortByPriority(ReadyList)*
 6:     **for all** *i ∈ ReadyList* **do**
 7:         **if** *i can be scheduled* **then**
 8:             *S(i) ← cycle*
 9:             *ReadyList ← ReadyList \ i*
10:         **end if**
11:     **end for**
12:     *cycle ← cycle + 1*
13:     *UpdateReadyInstructions(ReadyList)*
14: **end while**
15: **return** *S*

---

for instruction scheduling since the late 1970s for several reasons: it is easy to understand, can be easily adopted to different processor architectures, has a worst-case complexity of $O(n^2)$, and produces near-optimal results [CT04].

The basic idea behind list scheduling is sketched in Algorithm 5.2. The algorithm operates on a list of ready instructions (*ReadyList*), i.e., instructions whose predecessors in the dependence graph have already been scheduled. *ReadyList* is initialized with source nodes from $\mathcal{D}$, while the cycle counter is set to 0 (lines 1–2). Next, list scheduling iterates over *ReadyList* as long as there are unscheduled instructions (line 3). For each cycle, instructions in *ReadyList* are assigned a priority (line 4) which determines the order in which the instructions are scheduled by the greedy algorithm (lines 5–6).

Starting with instruction i having the highest priority, it is checked if i can be scheduled in the current cycle. Occupied functional units or results not yet generated by predecessor instructions are possible reasons why scheduling of i must be postponed. If i can be scheduled, it is added to schedule S (line 8) and removed from *ReadyList* (line 9). After processing all ready instructions, the cycle counter is incremented and the list of ready instructions is updated. Finally, the returned schedule S is used to reorder instructions in the original program.

One of the critical parts of Algorithm 5.2 is the priority assignment (line 4) that decides which instructions should be preferred among ready instructions. WCC's instruction scheduler combines different heuristics to compute the final instruction priority. To maximize superscalar execution, instructions that can be executed in a bundle are assigned a high priority. If more than one possible bundle exists, the number of immediate successors in the dependence graph $\mathcal{D}$ is computed for each instruction in the bundle. Bundles with a higher accumulated number of successors are preferred since an issue of such a bundle increases the number of instructions, which become ready, compared to issuing other instructions.

5.4.4 WCET-Aware Trace Scheduling

Trace scheduling is an extension of local scheduling. Instead of generating a good schedule for a single basic block, trace scheduling is a more aggressive approach that reorders instructions across multiple basic blocks. In the past, this global scheduling approach was studied in the context of an ACET reduction. Using profiling information, the run-time behavior of the program under analysis was employed to construct a trace that represents the most frequently executed path. In the next step, a good schedule was generated based on this trace.

The concept of traces was already presented in the course of the discussion about WCC's superblock optimizations (cf. Sect. 4.4 on p. 79). The techniques presented here resemble these previously presented optimizations in the sense that the trace selection is driven by WCET data. However, this time the trace is not constructed at source code but at assembly level. Moreover, the WCET-aware trace is not employed for superblock construction but is directly optimized via trace scheduling.

Shifting the focus from a single basic block to a sequence of blocks makes an extension of the dependence graph necessary. This is due to two reasons. First, conditional branches become scheduling candidates like all other instructions, thus, provisions in the dependence graph must be introduced that preserve the relative order of these instructions. Second, instructions may be moved across conditional branches which possibly leads to a changed semantics in off-trace paths. These two issues are formally expressed in the definition of the *trace dependence graph* $\mathcal{TD}$ for a trace (cf. Definition 4.3 on p. 83):

Definition 5.4 (Trace dependence graph) Given a trace $T = (b_a, \ldots, b_k)$, a *trace dependence graph* $\mathcal{TD} = (V_T, E_T)$ is a dependence graph defined on the node set $V_T = \{i \mid i \in \bigcup_{b \in T} b\}$. In addition to the data dependence edges, $\mathcal{TD}$ contains a set C of edges that model control flow dependencies between edges:

- Preserve order of branches:
$$\forall i_{branch}^1 \in T, |succ(i_{branch}^1)| > 1 : \forall i_{branch}^2 \in T, |succ(i_{branch}^2)| > 1 :$$
$$i_{branch}^1 <_T i_{branch}^2 \Leftrightarrow (i_{branch}^1, i_{branch}^2) \in C$$
- Preserve semantics in off-trace paths:
$$\forall i_{move} \in T : \forall i_{branch} \in T, |succ(i_{branch})| > 1, i_{branch} <_T i_{move} :$$
$$\exists i_{succ} \in succ(i_{branch}) : i_{succ} \notin T \wedge \mathrm{DEF}_{\mathrm{may}}(i_{move}) \cap \mathrm{LIVE\text{-}IN}_{\mathrm{may}}(i_{succ}) \neq 0)$$
$$\Leftrightarrow (i_{branch}, i_{move}) \in C$$

with $succ(i)$ being the set of successor instructions of i, and $(i_1, i_2) \in <_T \Leftrightarrow i_2$ follows i_1 in the sequence of trace instructions. $\mathrm{DEF}_{\mathrm{may}}(i)$ and $\mathrm{USE}_{\mathrm{may}}(i)$ represent def/use sets, i.e., sets of operands from/to which i may read/write.

Algorithm 5.3 Algorithm for WCET-aware trace scheduling

Input: $\mathcal{P}, timeout, updateWCEP, size$
Output: $(optimized)\ \mathcal{P}$
 1: $WCET_{org} \leftarrow WCETAnalysis(\mathcal{P})$
 2: $\hat{\mathcal{P}} \leftarrow \mathcal{P}$
 3: $counter \leftarrow 0$
 4: **repeat**
 5:    $T \leftarrow TraceSelection(\mathcal{P}, size)$
 6:    $MarkTraceBlocks(\mathcal{P}, T)$
 7:    $\mathcal{TD} \leftarrow BuildTraceDependenceGraph(T)$
 8:    $S \leftarrow ListScheduling(\mathcal{TD})$
 9:    $ReorderInstructions(\mathcal{P}, S)$
10:    $GenerateCompensationCode(\mathcal{P})$
11:    **if** $counter == updateWCEP$ **then**
12:       $WCET_{current} \leftarrow WCETAnalysis(\mathcal{P})$
13:       $counter \leftarrow 0$
14:    **else**
15:       $counter \leftarrow counter + 1$
16:    **end if**
17: **until** $all\ blocks \in \mathcal{P}\ scheduled$ **or** $exceeded(timeout) ==$ **true**
18: **if** $counter \neq 0$ **then**
19:    $WCET_{current} \leftarrow WCETAnalysis(\mathcal{P})$
20: **end if**
21: **if** $WCET_{org} > WCET_{current}$ **then**
22:    **return** $\mathcal{P}$
23: **else**
24:    **return** $\hat{\mathcal{P}}$
25: **end if**

Based on information from the trace dependence graph, WCC performs its WCET-aware trace scheduling. The main steps are shown in Algorithm 5.3 and will be explained in more detail in the following.

WCC's trace scheduler expects as input a program $\mathcal{P}$ and a timeout parameter used for the terminating condition. Moreover, parameter $updateWCEP$ specifies after how many trace scheduling iterations the WCET data should be updated by a timing analysis, while $size$ specifies the maximal number of blocks selected for the construction of a trace. In the beginning, a WCET analysis for $\mathcal{P}$ is performed and a copy $\hat{\mathcal{P}}$ of the original program is created (lines 1–2). Using this WCET information, WCC's rollback mechanism can be applied if $\mathcal{P}$ exhibits a WCET degradation after WCC's greedy trace scheduling. Moreover, the determined WCET data is used to select a WCET-aware trace (line 5) using WCC's longest path approach (cf. Sect. 4.4.3.1 on p. 84), while preserving size constraints. Blocks on the current trace are also marked in $\mathcal{P}$ to ensure that blocks that were already scheduled are excluded from further trace selection (line 6).

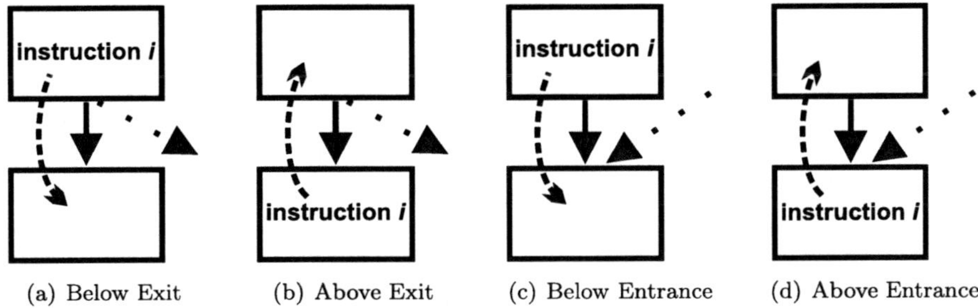

Fig. 5.9 Bookkeeping during trace scheduling

Using the identified trace, the trace dependence graph $\mathcal{TD}$ is constructed and passed to the list scheduler (cf. Algorithm 5.2) to find a schedule S that defines the new order of instructions (lines 7–9). Similar to Fisher's approach [Fis81], inner loops are represented on a trace as a single node, called *loop representative*, and code motion of instructions below or above these loops is allowed.

During scheduling, instructions could also move below or above a branch instruction. To ensure correct execution of the off-trace code, insertion of compensation code may be required (line 10). In literature, this process is referred to as *bookkeeping*. WCC uses standard approaches for compensation code generation which are explained in-depth in [Fis81]. For the sake of brevity, technical details are omitted here, and only a brief overview of the four possible scenarios that may occur (cf. Fig. 5.9), is provided:

(a) If instruction i is moved below a side exit (conditional branches out of the middle of the trace) and $\text{DEF}_{\text{may}}(i)$ is used before redefined when the side exit is taken, then copy i between the side exit and all uses of $\text{DEF}_{\text{may}}(i)$.
(b) Moving of i above a side exit (speculative execution) is only permitted if $\text{DEF}_{\text{may}}(i)$ is not live in the side exit. In this case, no compensation code is required.
(c) If instruction i is moved below a side entrance, a *rejoin point* (location in CFG behind new position of i where side entrance is moved to) must be identified and all instructions between the old side entrance and the rejoin point must be duplicated in a newly created basic block.
(d) If instruction i is moved above a side entrance, then copy i into all off-trace predecessor blocks of this side entrance.

After the generation of compensation code, Algorithm 5.3 determines the WCET of the scheduled program if necessary (lines 11–16). Otherwise, the scheduler possibly operates on outdated WCET data. However, this way the optimization run time can be reduced. WCC's trace scheduling is repeated as long as neither all blocks have been scheduled nor the timeout has exceeded (line 17). Finally, it is checked if WCET information is not obsolete (lines 18–20) and either the original or the optimized program is returned (lines 21–25). This mechanism takes care that WCC's greedy scheduler does not degrade the worst-case performance of the considered program.

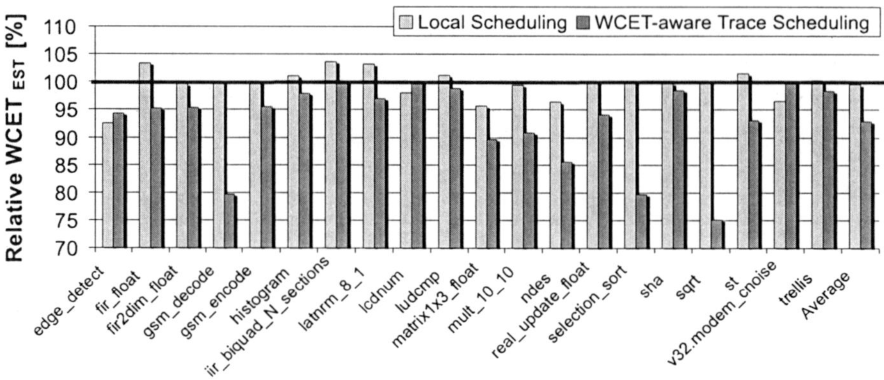

Fig. 5.10 Relative WCET estimates for WCET-aware trace scheduling

5.4.5 Experimental Results for WCET-Aware Trace Scheduling

In this section, the practical use of WCC's WCET-aware trace scheduling is demonstrated on real-life benchmarks. The experiments were performed on 20 representative benchmarks from the test suites DSPstone [ZVS+94], MediaBench [LPMS97], MiBench [GRE+01], MRTC [MWRG10], and UTDSP [UTD10].

For the experiments, the following optimization parameters were used: the timeout was set to 2 hours, the WCEP was updated via aiT after each second scheduling of a trace and the size of the selected traces was bounded to 100 basic blocks (cf. *updateWCEP* and *size* in Algorithm 5.3). These settings were also used for the WCET and ACET result diagrams presented in the following. Moreover, in a second scenario, the WCEP update was conducted after each 4th scheduling of a trace to compare the impact of a less frequent WCEP update on the WCET and optimization run time.

5.4.5.1 WCET

Figure 5.10 compares the impact of WCC's local and WCET-aware trace scheduling. The 100% base mark corresponds to the estimated WCET of the considered benchmarks compiled with WCC's highest optimization level *O3* and disabled instruction scheduling. The first bars per benchmark represent the results for local instruction scheduling. As can be seen, the improvements for most benchmarks are marginal. The best result is achieved for benchmark *edge_detect* with a WCET improvement of 7.6%. Some benchmarks, such as *fir_float* exhibit even a slight WCET degradation. A possible reason may be adverse line crossing effects (cf. Sect. 4.6.6). They can be introduced during scheduling when NOP instructions, which are generated in WCC's compiler backend as workaround for silicon bugs, are replaced by other instructions available in the code, leading to a modified memory layout. On average for the considered benchmarks, a WCET reduction of 0.4% was observed.

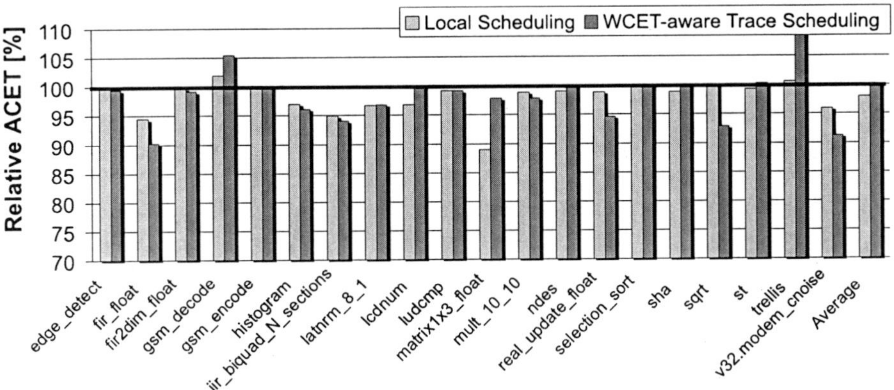

Fig. 5.11 Relative ACET for WCET-aware trace scheduling

Compared to local scheduling, WCET-aware trace scheduling yields significantly improved results. In most cases, trace scheduling benefits from the extended scope of multiple basic blocks and outperforms local scheduling. The best result was achieved for benchmark *sqrt* for which a reduction of the estimated WCET w.r.t. *O3* and local scheduling by 24.9% and 25.0%, respectively, was observed. Unlike local scheduling, WCET-aware trace scheduling never degrades the worst-case performance of a benchmark since any schedules leading to an increase of the WCET are discarded by WCC's rollback mechanism. On average for all benchmarks (see bar labeled with *Average*), trace scheduling outperforms *O3* and local scheduling by 7.1% and 6.7%, respectively.

5.4.5.2 ACET

WCC's trace scheduling focuses on most promising parts of the WCEP that constitute a trace. As discussed previously, these traces may not be consistent with the most frequently executed paths that are best suited for ACET reduction. Therefore, it is interesting to explore the impact of WCET-aware scheduling on the ACET.

The results are shown in Fig. 5.11, with 100% corresponding to the ACET for *O3* and disabled scheduling. Several conclusions can be drawn from this diagram. Most important, the positive impact of WCET-aware trace scheduling (second bars per benchmark) on the ACET is clearly smaller than on the estimated WCET. For example, optimizing benchmark *gsm_decode* yielded a WCET reduction of 20.3% but led to an increased ACET of 5.5%, translating to a difference between both metrics of 25.8%. In addition to *gsm_decode*, a degradation of the average-case performance was observed for another benchmark: the ACET of *trellis* increased by 43.0%. In contrast, such negative scenarios could be avoided for WCET reduction by using WCC's rollback mechanism.

On average, WCET-aware trace scheduling achieved an ACET reduction of merely 0.1%. This result is even worse than for local scheduling (see first bars per benchmark) that could reduce the ACET by 2.0% on average. Moreover, comparing the average results for WCET and ACET, it becomes clear that WCC's trace scheduler is tailored towards a WCET reduction since the performance improvements for WCET outperform the ACET reduction by 7.0%. This result conforms with previous observations in this book: optimizing code for the worst-case performance may be ineffective for an ACET optimization and vice versa.

5.4.5.3 Code Size

Scheduling instructions across branches may require insertion of compensation code. Even though it could be expected that the code size is significantly increased after trace scheduling, the results for the considered benchmarks only exhibit an average increase of 2.4%. The maximal code expansion was observed for benchmark *lcdnum* with 10.0%, while the increase is negligible for other benchmarks. The reason for this marginal increase is that no compensation code is required in many cases. Or only a small number of instructions must be duplicated which is a small fraction compared to the total code size. As a consequence, it can be concluded that trace scheduling is best suited for embedded systems with restricted memory resources that have to meet high run-time performance requirements.

5.4.5.4 Optimization Run Time

The improvement of the estimated WCET comes at the cost of an increased optimization run time. In a first scenario, which was also used for the aforementioned results, the WCET data was updated after two traces have been scheduled. WCC's WCET-aware trace scheduling of all benchmarks took a total of 107 CPU minutes, measured on an Intel Xeon 2.4 GHz system with 8 GB RAM. In contrast, the optimization run time for all benchmarks compiled with *O3* and local scheduling amounted to 31 minutes. Hence, it can be concluded that most time is spent for aiT's timing analyses that update the WCET data. The run time increase of more than 300% is fully acceptable if high code quality for embedded real-time systems is required.

In a second scenario, the interval for WCEP updates was increased—aiT was invoked after four traces have been scheduled. For this configuration, the WCET improvements were marginally decreased by 1.1%, while the optimization run time decreased by 11% to 96 minutes. These results conform the observations reported for WCC's superblock optimizations (cf. Sect. 4.4.5)—a too frequent WCEP update by a costly WCET analysis may not always pay off.

5.5 Summary

In this chapter, novel assembly level optimizations were presented that aim at a reduction of the WCET. Unlike numerous traditional ACET optimizations that transform code based on profiling information, WCC's optimizations are driven by WCET data. The availability of this timing information entails two main advantages. On the one hand, it allows a systematic improvement of the code lying on the WCEP. On the other hand, any modification of the code can be quantified with respect to the program's worst-case behavior. Moreover, the presented WCC assembly level optimizations exploit processor-specific knowledge about the underlying architecture in order to generate high quality code that is tailored towards the TriCore processor. The discussed optimizations were integrated into the WCC framework and their effectiveness was demonstrated on real-life benchmarks.

The first optimization presented in this chapter was procedure positioning. Its goal is the improvement of the I-cache behavior by re-arranging procedures in memory in such a way that mutual cache evictions are minimized. Unlike traditional profiling-based positioning, WCC's WCET-aware positioning uses worst-case call frequencies to construct a call graph that reflects call relationships between procedures on the WCEP. The optimization was implemented as an iterative, greedy approach that evaluates each modification of the memory layout as well as a fast heuristic approach. For the considered benchmarks, the greedy algorithm achieved an average WCET reduction of 10.1%, while the heuristic approach reduced the WCET by 3.1% on average.

The second optimization discussed in this chapter was trace scheduling which is a global instruction scheduling technique. In contrast to local scheduling, which confines its attention to a single basic block, trace scheduling operates on a sequence of basic blocks called trace. This way, more optimization opportunities are provided and a higher instruction-level parallelism can be exposed. Contrary to ACET scheduling relying on profiling data, WCC's WCET-aware trace scheduling identifies a trace by exploiting WCEP information. Using these traces, an instruction order is generated that fully utilizes the superscalarity of the TriCore processor. Experiments on real-life benchmarks show that on average WCC's trace scheduling could outperform local scheduling by 6.7% which translates to an average reduction of the WCET w.r.t. the highly optimized code without scheduling of 7.1%.

The results in this chapter point out two important issues. First, it is crucial to use a specific metric during optimization. If high WCET reduction is the primary goal, code transformations should be driven by WCET data and not by profiling information which may classify other paths than the WCEP as critical. Using this WCET data allows a systematic improvement of the WCEP even at the cost of other paths. Second, the exploitation of processor-specific features during code transformation may yield a significant improvement of the program performance. Hence, high code quality for embedded systems can only be achieved if an optimizing compiler applies aggressive code transformations at both the source code and assembly level.

Chapter 6
Machine Learning Techniques in Compiler Design

Contents

6.1 Introduction

The previous sections have demonstrated that the development of heuristics for compiler optimizations which efficiently reduce a given objective, such as the WCET estimation, is a tedious task. This process requires both a high amount of expertise and an extensive trial-and-error tuning. However, even though the previously presented optimizations utilize complex analyses and code transformations, it becomes clear that their transformations are often not optimal—one indicator for the non-optimality is the common phenomenon of performance degradations after a code transformation.

P. Lokuciejewski, P. Marwedel, *Worst-Case Execution Time Aware* 159
Compilation Techniques for Real-Time Systems, Embedded Systems,
DOI 10.1007/978-90-481-9929-7_6, © Springer Science+Business Media B.V. 2011

The two main reasons why compilers fail to deliver optimal performance are the following:

- Compiler optimizations have shown to be the source of some of the hardest and most challenging problems in computer science [SS07]. Since finding optimal solutions to many optimization problems is provably hard, compiler writers are forced to use heuristics as approximate solutions. However, heuristics are often based on abstract models of the underlying complex computer architectures that are not capable of precisely capturing all relevant architectural features. Using these inaccurate abstractions leads to a set of universal compiler heuristics that do not exploit a particular target architecture and might even have a negative impact on the program to be optimized.
- In typical compilers, optimizations are not performed separately but within a sequence of interfering optimizations. Since optimizations may have conflicting goals, disadvantageous interactions can be observed. However, formal models expressing interactions between optimizations are hardly known. As a consequence, compiler writers are forced to either tune a heuristic for a given optimization sequence that must not be changed, or their heuristics are based on conservative assumptions which do not allow the exploration of the program's optimization potential.

A solution to this dilemma is the application of *machine learning* (*ML*) approaches to automatically generate heuristics. The key advantage of these techniques is their capability to find relevant information in a high dimensional space, thus helping to understand and to control a complex system. Providing a set of characteristics, so-called *static features*, about the code to be optimized, a machine learning algorithm automatically learns a mapping from these features to heuristic parameters.

Using the learning result enhances the flexibility of a compiler framework. Exchanging the target architecture or modifying the optimization sequence requires only an automatic re-learning to adjust heuristics. Thus, machine learning can be leveraged to reduce the efforts of developing compiler optimizations, an issue which is crucial in today's rapidly evolving processor market. Another advantage of these heuristics is their performance as machine learning based (MLB) heuristics may often outperform hand-crafted heuristics [MBQ02, SAMO03, CM04].

Up to now, machine learning techniques have been studied in the context of ACET optimizations. This book is the first to exploit machine learning techniques for WCET reduction.

In this chapter, an automatic generation of MLB heuristics for WCET-aware compiler optimizations, which was developed in this book, is presented. Section 6.2 provides an overview of research that pursues a reconciliation of statistical and machine learning techniques with compiler optimizations. In Sect. 6.3, the general workflow of machine learning based heuristic generation is introduced. A feasibility study for the generation of WCET-aware heuristics for the source code optimization function inlining is presented in Sect. 6.4. Section 6.5 goes even one step further. On the one hand, this section shows that MLB WCET-aware heuristics can also be easily constructed at other abstraction levels of the code, namely at assembly level as demonstrated for the optimization loop-invariant code motion. On the other hand, this section indicates that an automatic selection of appropriate machine learning algorithms and their parameters can significantly improve the performance of the generated heuristics. Finally, this chapter is concluded in Sect. 6.6.

6.2 Related Work

MLB Heuristic Generation for ACET Reduction   The automatic generation of optimization heuristics has been studied for different ACET optimizations. Monsifrot et al. [MBQ02] used a supervised classification to generate heuristics for loop unrolling which decide whether unrolling should be performed or not. This two-class classification problem was extended by Stephenson [SA05] to find MLB heuristics that predict the best unrolling factor for a given loop. Machine learning techniques were also used for instruction scheduling. Russell et al. [RMC+09] used supervised learning to induce heuristics for superblock instruction scheduling. In [CM04], supervised learning was employed to generate basic block scheduling heuristics for a Java just-in-time (JIT) compiler to tackle the trade-off between optimization run time and program performance by applying scheduling only to those blocks that promise a significant improvement. Calder et al. [CGJ+97] exploited neural networks and decision trees to predict branch behavior based on static features associated with each branch.

In [LBO09], a grammar-based mechanism using genetic programming was presented that automatically extracts features from the internal code representation of GCC for MLB heuristics. Genetic programming was also utilized for the generation of compiler heuristics for the optimizations hyperblock formation, register allocation, and data prefetching [SAMO03].

Other Application Fields of Machine Learning   Besides heuristic generation, machine learning can help to characterize mutual interactions between optimizations as well as interactions between optimizations and the architecture. For example, Vaswani [VTS+07] used empirical regression models to characterize interactions between optimizations in GCC. The drawback of this approach is that each optimization and the used hardware platform must be formally specified for the regression model. For arbitrary transformations and processors, this requirement is hard to achieve.

Machine learning can also be applied to determine promising settings for optimization flag and parameters. Haneda et al. [HKW05] exploited non-parametric statistical analysis to decide which GCC compiler flags should be turned on or off. Cavazos et al. [CO06] used logistic regression to automatically derive a model that determines which optimizations of a Java JIT compiler should be applied to a function based on its static features.

6.3 Machine Learning Based Heuristic Generation

Machine learning is a scientific discipline that can be subdivided into different classes. The most popular classes are *supervised* and *unsupervised learning* as well as *reinforcement learning* [Bis08].

Unsupervised learning determines how data is organized and allows its clustering. For the generation of compiler heuristics, which have to decide if a particular optimization should be applied or not, unsupervised learning is usually not applicable. Reinforcement learning uses an agent that tries different actions and observes their consequences that are expressed as rewards. Based on this knowledge, the agent learns how to act in different situations, e.g., whether a code optimization should be performed. In this book, the compiler heuristics are generated by means of supervised learning, a technique that tends to be faster than reinforcement learning for heuristic generation [SA05]. Moreover, only binary classification, which decides whether a particular optimization step should be performed or not, is considered.

6.3.1 Supervised Learning

Heuristic generation can be reduced to the problem of supervised classification. The fundamental idea behind this machine learning based approach is to consider a compiler heuristic as a function that takes different characteristics describing the situation found by the optimization as input and computes an output that reflects the performance of each optimization step. This reduction enables the application of standard machine learning techniques.

Classifier learning is performed on observations x_i of an object i, with $i = 1, \ldots, N$, that were gathered in the past, together with their binary classification (or *label*) $y_i \in \{0, 1\}$, where 0 can be mapped to *NO* and 1 to *YES*. Each observation x_i represents a vector of p features $x_i = (x_i^1, \ldots, x_i^p)$, with $p = 1, \ldots, M$. Due to the given classifications, this learning task is called *supervised*. A (learning) *example* is a pair $\langle x_i, y_i \rangle$. The set of N examples used for learning is called the *training set*. Formally, the supervised learning problem can be described as follows:

Definition 6.1 (Supervised learning) Given a set of possible observations X, a set of output labels Y, and a training set $E = \{(x_1, y_1), (x_2, y_2), \ldots, (x_n, y_n)\}$, where $x_i \in X, y_i \in Y$. *Supervised learning* is a machine learning technique that finds a function $f : X \to Y$ such that a given performance function *EVAL* is optimized.

A possible performance function is *accuracy* which requires another set of examples, the *test set*, that was not used in the learning phase. The learned classifier f is applied to observations from the test set and the computed output labels are then compared to the true labels. If the true label and the one predicted by the learned classifier are the same, the observation was classified correctly, otherwise it is an error. The accuracy of a supervised learner is the ratio of all correctly classified observations divided by the total number of observations in the test set.

For the generation of compiler heuristics, an object i represents an optimization candidate, such as a function or an instruction, that is characterized by the feature vector x_i. Labels express the effect of optimizing i for a particular cost function. For example, a reduced program's WCET after optimizing i may be labeled as *YES*, otherwise *NO*.

6.3.2 Heuristic Generation Based on Supervised Learning

The heuristic generation begins with the obvious decision for which compiler optimization an improved heuristic should be generated. Afterwards, the following steps as depicted in Fig. 6.1 are conducted:

- **Representation of the Program**
  The first step is a representation of the program under test by internal compiler data structures. Examples are high- or low-level intermediate representations, abstract syntax trees, or control flow graphs.
- **Feature Extraction/Label Determination**
  Based on the program representation, the developer has to decide which features best characterize the parts of the program to be optimized and how they can be extracted. Possible features are the number of successors of a given basic block or the number of floating point instructions within a block. The features must be transformed into a proper vector representation, serving as input for the ML tool. This process is called feature extraction. In addition, for each feature vector a label representing the desired output has to be determined. The result of this step is a constructed training set.
- **Machine Learning Algorithm/Parameter Selection**
  The next step in Fig. 6.1 is the selection of a supervised learner and its parameters. The machine learning community has developed a large portfolio of different learners over the last decades. Moreover, many learners have several user-defined parameters, leading to models with different performance. Due to the large number of possible combinations, the selection of the appropriate learning algorithm is not straightforward.
- **Model Induction:**
  The chosen classifier (learner) induces a prediction model representing a compiler heuristic which can be used to predict if/how the considered optimization should be performed for unseen data.

6.3.3 Integration and Use of MLB Heuristics in Compilers

Depending on the type of the induced model, the MLB heuristics can be in integrated in two different ways into the compiler. If the learning algorithm produces a human readable model, then the heuristic can be directly integrated into the optimizer of the compiler. For example, an induced decision tree can be implemented as nested

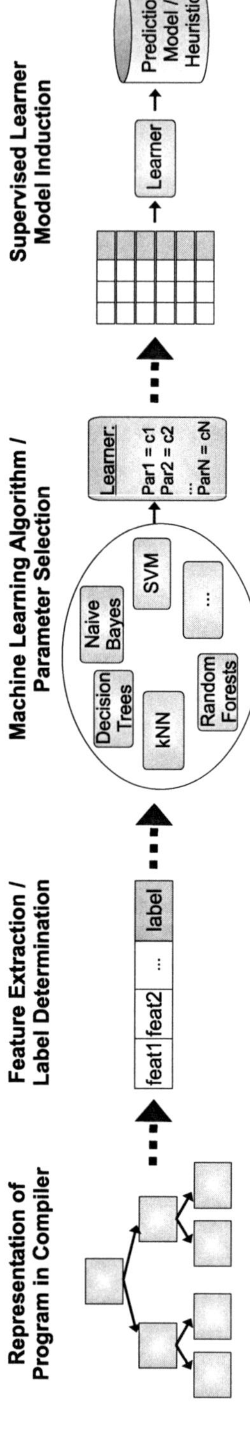

Fig. 6.1 Overview of machine learning based compiler heuristic generation

`if-then-else` constructs. The advantage of such a model representation is that the heuristic can be compiled and shipped with the compiler without the need of any external tools. However, many learning algorithms produce models that can not be simply translated into programming language constructs, thus a direct integration into the compiler source code is often not feasible.

For the latter class of algorithms, an efficient communication between the compiler and the machine learning tool is mandatory. Whenever an optimization decision based on an MLB heuristic has to be taken, the compiler computes the feature vector and hands it to the ML tool. The tool uses this data to predict a class and sends the result back to the compiler. Such an integration of machine learning into a compiler framework may introduce some overhead during the optimization. However, if the implementation is efficient, the overall overhead is small. On the other hand, a reconciliation of a compiler with an ML tool offers a compiler designer a vast spectrum of supported ML techniques for optimizations and analyses.

6.4 Function Inlining

Function inlining is a well-known transformation replacing the function call by the body of the callee while storing the arguments in variables that correspond to function parameters. Due to the reduced calling overhead and the capability of function inlining to enable other optimizations, most modern compilers apply this transformation for a reduced ACET. However, the introduced structural changes of the code after the optimization entail hardly predictable interferences with other optimizations and the target architecture. Therefore, the generation of effective inlining heuristics is challenging.

In this section, it is demonstrated how supervised learning can be utilized for WCET reduction using the source code optimization function inlining. After a motivation in Sect. 6.4.1, which indicates possible adverse effects of inlining when simple heuristics are employed, Sect. 6.4.2 introduces standard function inlining in more detail and points out the positive and negative effects of the transformation on the program performance that may occur during inlining. To improve standard inlining heuristics, MLB WCET-aware inlining heuristics are constructed in Sect. 6.4.3 and their predominance compared to standard inlining is presented on real-life benchmarks in Sect. 6.4.4.

6.4.1 Motivating Example

The decision if a particular function should be inlined is made by compiler heuristics. They try to predict if the application of the optimization will be beneficial. The most common heuristic found in literature [BGS94, Muc97] and many compilers is the consideration of the callee size. This parameter can be expressed as the number of source code expressions or assembly instructions. If a predefined threshold

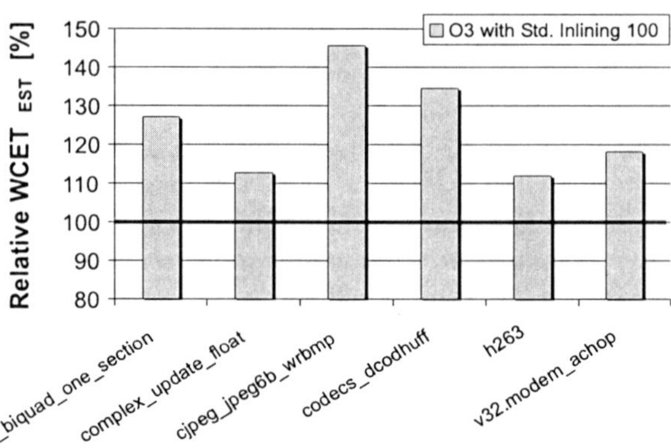

Fig. 6.2 Negative impact of function inlining on WCET

is exceeded, function inlining is omitted. Usually, this heuristic is conservative and allows inlining of only small functions to avoid too severe effects.

However, due to the complex interactions between function inlining and other optimizations as well as the architecture, a simple heuristic based on the callee size is not sufficient for an effective run-time optimization. Even worse, the simple heuristic may inline inappropriate functions which substantially degrade the system performance.

The negative influence of WCC's standard function inlining on selected real-world benchmarks is illustrated in Fig. 6.2. The bars represent relative WCET estimations for the code compiled with the highest optimization level (*O3*) and enabled inlining using the simple heuristic based on the callee size which is bounded to 100 expressions. 100% corresponds to the WCET for *O3* with disabled inlining. The code is executed from TriCore's cached Flash memory.

It can be seen that the WCET was increased by up to 45.7% for the benchmark *cjpeg_jpeg6b_wrbmp* compared to the non-inlined version. Thus, the figure emphasizes that the optimization might adversely affect the program.

Hence, it is important to correctly decide whether to apply function inlining or not. One possible solution to find a more accurate heuristic is to invest a lot of effort into an empirical evaluation of different inlining decisions. Another solution, that is discussed in the next section, is an automatic generation of inlining heuristics by machine learning techniques that demand a fraction of effort compared to the hand-crafted heuristic generation and still produce comparable or even better results. The task is then to extract features that characterize a function call, thus establishing the observations to choose an appropriate learning algorithm, and to integrate the learning result into the compiler.

6.4.2 Standard Function Inlining

Function inlining replaces function call sites by the body of the corresponding callee. This so-called *inline expansion* can be either performed by the compiler in the high-level or low-level IR, or by the linker. Typically, inlining is applied to reduce the ACET due to the following reasons:

- By copying the callee's function body into the caller, the calling overhead is reduced since function call and return instructions as well as the parameter handling is removed.
- Removing calls and returns reduces control transfers, thus improving the pipeline behavior.
- Inlining enables other optimizations that could not be applied to the original code since they were restricted by function boundaries.

The evaluation of the impact of function inlining on the program execution is challenging since the transformation influences different components, e.g., register allocation, instruction scheduling, and the usage of the memory hierarchy. Thus, its consequences are not directly visible but become apparent as side effects. This complicates the decision of whether a function should be inlined. Although it is widely believed that function inlining substantially improves the program run time, different studies, such as [CHT91] as well as the previously presented experiments, came to the conclusion that its application is not always beneficial.

One of the main drawbacks of this optimization is the increased register pressure. By inserting additional variables from the inlined function into the caller, possibly more registers are required. If the callee is inserted in an area with an already high register pressure, the register allocator is afterwards forced to add spill code. These additional accesses to the memory degrade the performance. Another problem that function inlining entails is a possibly degraded I-cache behavior. With an increasing code size, run-time critical code may not remain in the cache but are evicted by the inlined function. The resulting cache conflict misses slow down the program execution. Furthermore, cache performance may suffer from inlining since the locality of references is decreased.

6.4.2.1 Related Work

The influence of the optimization function inlining on the ACET has been studied in different publications. In [DH89], equations representing the performance of inlined versions of the programs have been formulated, revealing which factors affect the performance of inlined code. The effectiveness of inlining has also been evaluated in [CHT91]. From their experiments, the authors of both works concluded that this optimization was not always beneficial for the program performance.

Inlining was also studied in the context of evolutionary algorithms. Cavazos [CO05] used genetic algorithms to tune the parameters of inlining heuristics in a dynamic Java compiler. This publication is close to the work presented in Sect. 6.4.3, but also differs in several important ways. Most importantly, the goal of WCC's WCET-aware inlining is the reduction of the worst-case and not of the average-case performance. Furthermore, Cavazos employed a genetic algorithm, while WCC utilizes supervised learning of a classifier.

6.4.3 MLB Heuristic Generation at Source Code Level

In this section, an automatic generation of MLB heuristics for the optimization
WCET-aware function inlining applied at source code level is proposed. The heuristic generation follows the scheme depicted in Fig. 6.1. Moreover, it is shown how
the novel heuristic is integrated into WCC and how inlining is applied to achieve a
high WCET reduction while keeping the code expansion as small as possible.

6.4.3.1 Feature Extraction

The first step is the construction of a training set. For this purpose, examples must
be collected that appropriately characterize the inlining candidates, i.e., callees that
may be inlined.

The feature extraction exploits WCC's infrastructure. It gathers data from the
high-level intermediate representation ICD-C IR and uses the back-annotation to
obtain information from the compiler backend, such as WCET-related data computed by aiT. In total, 22 features are extracted that are considered to be most significant for the classification. Numerical features have in general a value range of $\mathbb{N}_0$
and are either expressed as absolute numbers or as percentages (*relative features*).
Among others, the following numerical and binary features are extracted:

- size of caller/callee measured in number of expressions
- number of call sites of the callee in caller
- worst-case execution frequency of inlining candidate in caller
- total WCET of caller/callee
- WCET of a callee for a particular call site, i.e., the product of the callee's WCET
  and the worst-case execution frequency of that call site
- relative WCET of a caller/callee w.r.t. the overall program WCET
- whether a callee is invoked from exactly one call site
- whether the call site is located in a loop

Register Pressure Analyzer   As discussed in Sect. 6.4.2, function inlining can
potentially increase the register pressure that is defined as follows:

Definition 6.2 (Register pressure) Let θ be a program control point in the control
flow graph of program $\mathcal{P}$. The *register pressure* is the number of temporaries, such
as virtual registers, that are live at θ. If the underlying architecture model supports
separate register sets, the register pressure is computed separately for each set.

During the invocation of a function, *caller-saved* registers are usually saved before the callee is entered and restored after returning to the caller. This so-called
context save enables the usage of saved registers in the callee. Performing function
inlining and replacing the call site by the callee's function body makes the context

save redundant. However, as a consequence, the number of registers that can be exploited in the inlined function is decreased, leading to a higher register pressure in the caller.

Increasing the register pressure can potentially lead to a generation of spill code which has a negative effect on the program run time. Ideally, a WCET-aware inlining heuristic should be able to predict whether inlining a function yields spill code in order to prevent the application of this optimization and to anticipate an increase in the WCET. For this purpose, the *register pressure analyzer* (*RPA*) was developed in the WCC framework.

The RPA takes those registers into account that can be potentially spilled, i.e., all address and data registers of the TriCore processor for which the compiler can produce spill code. Input to WCC's RPA is the physical LLIR code (cf. Sect. 3.3 on p. 31) representing the program under analysis. Information about the liveness of registers is computed by a lifetime analysis [App97].

For feature extraction, the RPA provides significant information concerning the classification of inlining candidates based on the register allocation. The following additional features are taken into account:

- number of address/data registers whose lifetimes span a call. Inlining of calls crossing a high number of lifetimes increases the probability for spilling.
- maximal number of registers that are simultaneously live within a function. Inlined callees with a high register pressure are more likely to introduce spilling.

6.4.3.2 Label Determination

Besides the feature extraction, each example of the training set must be assigned a label. The labels in the training set indicate if inlining a function specified by its feature vector was successful concerning the WCET reduction.

The labels are automatically determined. For this purpose, WCC is run twice. In a first run, the analyzed program is compiled with the highest optimization level (*O3*) and disabled standard function inlining. Next, the overall program WCET $WCET_{ref}$ is computed and serves as reference value. In a second run, the same program is compiled again with *O3* and function inlining disabled for all function calls except for the call site i which is the currently considered inlining candidate. For this generated code where exclusively i was inlined (independent of the callee's size), the program's WCET $WCET_i$ is determined. A comparison between both WCETs indicates the influence of inlining call site i on the estimated WCET.

The binary value of $label_i$ for call site i is determined as follows:

$$label_i = \begin{cases} yes, & \text{if } \frac{WCET_i \cdot 100}{WCET_{ref}}\% \leq 99\% \\ no, & \text{otherwise} \end{cases}$$

The value of 100% means that inlining of function call i had no influence on the WCET. If the value is less than 100%, a WCET reduction was achieved, otherwise inlining had a negative impact on the worst-case performance. A threshold of 99% is used, i.e., if inlining at call site i reduced the program's WCET by at least 1%, this call is considered as being beneficial. This strategy is motivated by the potential code expansion which is a crucial issue in particular for embedded systems. The threshold ensures that a minimal WCET reduction of less than one percent coming with a potentially large code size increase is not classified as a positive inlining example.

The label extraction is automatically performed for all call sites in the benchmarks of the training set that allow function inlining. For each call, in a first step its features are extracted and in a second step, the corresponding label is determined. These examples are collected for all considered benchmarks and serve as input to the automatic generation of a heuristic.

6.4.3.3 Model Induction

There are several algorithms for supervised learning, leading to the question which learner should be employed for the heuristic generation. For the feasibility study of MLB WCET-aware function inlining, the following requirements are desirable:

- highly precise classification
- understandability of the induced model
- easy integration of the model into the compiler

While linear models, which determine a hyperplane in the feature space separating positive and negative examples, often deliver highly accurate results even in high-dimensional spaces, their learning result is hard to understand.

In contrast, partitional approaches like decision trees can be easily interpreted. A *decision tree* splits the set of examples according to the values of one feature (a vector component). In each subspace, this step is recursively repeated until all examples in a subspace belong to the same class. In other words, a decision tree consists of a sequence of decisions represented by a tree. Each node in the tree performs a test on the values of a single or several features, splitting the path based on possible outputs. Leaves in decision trees represent the final output, which is the predicted class. To classify an object, the tree is traversed on a particular path from the root node to a leaf, depending on the test results that are based on the values of the feature vector.

Ensemble techniques [Bis08] which combine several learning results and classify according to the majority of the predicted classifications have shown to induce models of high accuracy. Hence, an ensemble of decision trees is a qualified approach for the feasibility study of MLB WCET-aware function inlining presented in this book: it does not only promise a precise classification but the results are also human readable and can be converted into an `if-then-else` structure that can be implemented into WCC's optimizer.

Random Forests A popular ensemble technique for decision trees is *random forests* [Bre01]. Random forests consist of many *unpruned* decision trees constructed from different *bootstrap* samples which are obtained from a training set D of size N by sampling examples from D uniformly and with replacement. In contrast to standard trees where node splits are based on all features from the training set, random forests use a randomly chosen subset of features to find the best split for each node.

This counterintuitive strategy turned out to be highly accurate and generates results comparable with other state-of-the-art algorithms like *Support Vector Machines* or *Adaboost* [Bis08]. Other important advantages of random forests are their speed, their robustness against overfitting (i.e., the model is too sensitive to characteristics of training set used to build the model), and their user-friendliness since just two parameters (number of considered features for node splitting and number of trees in forest) have to be defined.

To predict novel data, the object under consideration is classified by each of the trees in the forest and the outputs are aggregated by a *majority vote*, i.e., the most frequently predicted class is the final output.

6.4.3.4 Application of WCET-Aware Function Inlining

One of the advantages of random forests is the possibility of transforming the classification rules into equivalent programming language constructs. Since random forests are a collection of decision trees representing tests of conditions concerning the features, they can be translated into `if-then-else` statements that are incorporated into WCC as optimization heuristics.

To achieve a high WCET reduction simultaneously with a small increase of code size, WCC combines the MLB heuristics with a complementary code-size related heuristic for the selection of inlining candidates. Ordinary compilers typically consider inlining candidates in a top-down manner by traversing the program code. In contrast, WCC starts with the optimization of callees having the largest worst-case execution counts (WCECs). These are usually functions that promise the largest WCET reduction due to subsequently applied optimizations that follow inlined. This way, WCC exploits high optimization potential with the minimal number of transformations, avoiding a too severe code expansion in the first optimization steps. After finding such an inlining candidate with maximal WCEC, the novel MLB heuristic is applied to decide if function inlining should be performed.

If it was decided to inline the considered function, a WCET analysis must precede the next feature extraction in order to update WCET-related data. Otherwise, the previously computed WCET information can be reused from the previous timing analysis. Since in many cases inlining is prevented, the required number of WCET analyses remains small. Compared to a conventional WCET optimization that would evaluate the impact of each inlining step via a timing analysis, WCC's classifier can considerably cut down the optimization run time.

6.4.4 Experimental Results for WCET-Aware Function Inlining

To indicate the efficacy of WCC's WCET-aware function inlining, an evaluation on a large number of different benchmarks was performed. In total, 41 benchmarks were used for the training set and 40 benchmarks for the (disjunct) test set. The benchmarks come from the test suites DSPstone [ZVS+94], MRTC [MWRG10], Media-Bench [LPMS97], MiBench [GRE+01], NetBench [MMSH01], UTDSP [UTD10], and WCC's own collection of real-life benchmarks representing miscellaneous applications, e.g., an H263 coder or a G.721 encoder. The code size of the benchmark codes ranges between 58 and 17,864 bytes.

The tests during the training phase are conducted using two different types of memories available in the TriCore TC1796 processor. The first memory is the program scratchpad memory (SPM) with a capacity of 48 Kbyte, the second memory utilized for the experiments is a 1 MByte cached program Flash memory. To take cache effects, which are crucial for the code expansion during inlining, into account, the cache capacity for the WCET analysis was reduced from 16 KByte to 2 KByte.

Due to architectural reasons, the system behavior may differ depending on whether the program is executed from the SPM or Flash. Thus, two different training sets for each memory type were generated, each of which contains 275 feature vectors that were extracted from all benchmarks, with approximately 70% of *negative examples*, i.e., function inlining had a negative effect on the WCET for these cases. This number points out that function inlining is an optimization that should be used with caution. The number of extracted examples ranges from 1 example for small benchmarks up to 59 examples for larger benchmarks like *g721_encode*.

WCC's feature extraction for function inlining takes place at source code level in the ICD-C IR. Additional features concerning the WCET and code size are provided by WCC's back-annotation. Finally, information about the liveness of registers are generated by WCC's register pressure analyzer (cf. Sect. 6.4.3) and back-annotated into the compiler frontend.

The constructed training set was processed by the open-source machine learning tool *RapidMiner* [MWK+06] which induced the classification model based on the learner random forests. For the learner, the following default parameters were selected: number of decision trees was 13, number of considered features at each split was $\log(i)$ (with $i = 22$ being the total number of considered features), and the splitting criterion was the *gini index* [Bis08].

6.4.4.1 Accuracy of Classification

The quality of a classifier can be estimated by the accuracy which indicates how well the classification model will perform in the future for unseen examples. To evaluate WCC's random forests heuristic, the common approach called *leave-one-out cross validation (LOOCV)* [Bis08] is applied. LOOCV eliminates a single example from the training set of size n, exploits the remaining $n - 1$ examples to learn a classifier and finally uses the eliminated example to validate the classification model.

Table 6.1 Accuracy and class recall based on LOOCV

	SPM	Flash
Correctly classified examples (accuracy)	84.0%	83.5%
Correctly classified positive examples	36.1%	19.2%
Correctly classified negative examples	97.7%	99.5%

This validation is repeated n times for each example considered once as test set. The estimated performance of a learner is the average of the measurements of all n runs. LOOCV is best suited for applications with only a small number of available learning examples since the learning algorithm can be precisely trained with almost all examples [SA05].

Table 6.1 summarizes the LOOCV results for the training set constructed for the SPM and Flash memory. The second row represents the accuracy for all examples of the training set. The last two rows indicate the so-called *class recall* for the positive and negative examples which are examples with the labels *YES* and *NO*, respectively.

In total, 84.0% and 83.5% of the examples could be correctly classified for SPM and the Flash memory, respectively. It can be observed that the classification error for positive examples (classifier predicts inlining as beneficial) is significantly larger than the error for the negative examples. A reason for that is the small number of positive examples used in the experiments. Thus, the learning algorithm was not able to generate an accurate model for positive examples as was accomplished for the negative examples. In general, these results are promising for WCC's WCET-aware function inlining since it is very likely that inlining decisions leading to a degraded worst-case performance will be predicted and thus avoided.

6.4.4.2 Variable Importance Measure

One advantage of random forests is their capability of estimating the importance of features (aka. variables) for the classification, called *impurity measures*. The importance of a feature is determined by its contribution for an effective classification and reveals if the features chosen for the learning algorithm are appropriate. Figure 6.3 depicts the importance of attributes discussed in Sect. 6.4.3 which is based on the gini index for the Flash data set. The higher the gini index for a feature, the more important it was for the classification.

As expected, the most important variable for the classification of inlining candidates is the size of the callee. This is also the attribute that is found in inlining heuristics of most compilers. However, as could be seen for the WCET results, the exclusive consideration of this attribute is often not sufficient and might lead to an inappropriate inlining decision.

It can also be seen that attributes concerning the program's worst-case behavior are highly important. This underlines that an optimization that is tailored towards a WCET reduction should take WCET information during the learning phase into

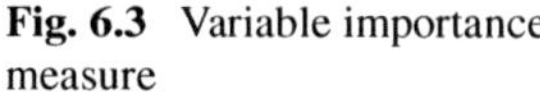

Fig. 6.3 Variable importance measure

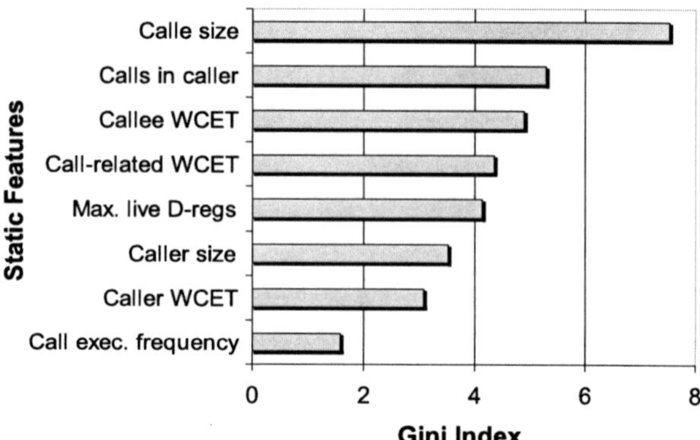

Table 6.2 Training set: overview of the impact on WCET

	WCET$_{EST}$ - SPM	WCET$_{EST}$ - Flash
Std. Inlining 50	101.7%	100.3%
Std. Inlining 100	104.6%	105.5%
MLB WCET Inlining	92.6%	94.1%

account. Last but not least, Fig. 6.3 reveals that the attribute characterizing the number of simultaneously live data registers (labeled with *Max. live D-regs*), being an indication for potential spill code, is important. Thus, a register pressure analyzer is a source of crucial information when dealing with code-size critical optimizations.

6.4.4.3 WCET

In order to evaluate the effectiveness of WCC's machine learning based (MLB) inlining heuristic and to compare it against the standard ACET inlining heuristics, the WCET for real-life benchmarks after the code transformation was estimated.

In a first run, a *resubstitution estimation* was performed. It uses the entire training set to build a classifier and to estimate the performance of this model. This performance estimation is usually optimistic compared to the true error rate since the model is not tested on data that it has not already seen. However, it is often used as a first indicator which shows how well a particular learning algorithm could cope with the given training data. If a low performance of the learner, i.e., a high error rate, is observed, it is likely that the selected learner is not appropriate for the considered problem.

Table 6.2 shows the overall results achieved for the two memory types SPM and the cached Flash. The reference value of 100% is the WCET estimation of the program compiled with the highest optimization level (*O3*) and disabled function inlining. In the second row, a standard inlining heuristic was used (optimizing functions smaller than 50 expressions) for the estimation of the relative WCET (average

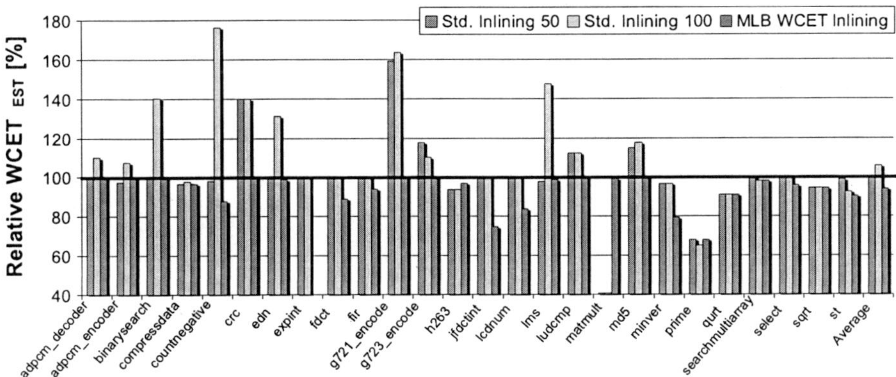

Fig. 6.4 Relative WCET estimates for WCET-aware inlining (training set)

for all benchmarks). As comparison, in the third row the standard inlining heuristic was increased to 100 expressions. Finally, the last row represents the relative WCETs when the standard heuristic is replaced by the MLB WCET-aware inlining heuristic.

The results point out that WCC's new heuristic outperforms the standard hand-crafted heuristic (limiting callee size to 50 expressions) on average by 9.1% for SPM and by 6.2% for the Flash memory. When the standard inlining heuristic is increased to 100 expressions for the inlining candidates, WCC's MLB inlining outperforms standard inlining on average by 12% for the SPM and 11.4% for the Flash memory. Moreover, comparing rows two and three in Table 6.2, the conclusion can be drawn that it is more beneficial to have a conservative heuristic that prefers optimization of smaller functions since inlining of larger function might have a strong negative effect on the program run time.

A comparison between the `if-then-else` constructs representing the SPM and Flash MLB heuristic, respectively, reveals that the models are similar showing only slight deviations. These deviations are a result of different labels which influence the decision tree generation. The differences in the labels are due to the fact that inlining candidates characterized by equal feature vectors may exhibit a different impact for the SPM or cached Flash memory. Due to the similarities of both heuristics, the following results will be presented for the Flash memory which is the commonly used storage type in real applications.

Figure 6.4 shows detailed resubstitution results for a selected subset of 26 representative benchmarks from the training set, omitting benchmarks for which comparable results were achieved for the MLB and standard heuristic. The bars represent relative WCET estimations for Flash memory, with 100% corresponding to the WCET for *O3* and disabled inlining. The bar labeled with *Average* indicates the average WCET results achieved for all 41 benchmarks from the training set. As can be seen, the MLB heuristic never significantly increases the WCET for all benchmarks compared to standard inlining.

Three benchmarks are discussed in more detail which summarize the typical impact of the new heuristic. The benchmark *crc* benefits from the new heuristic since

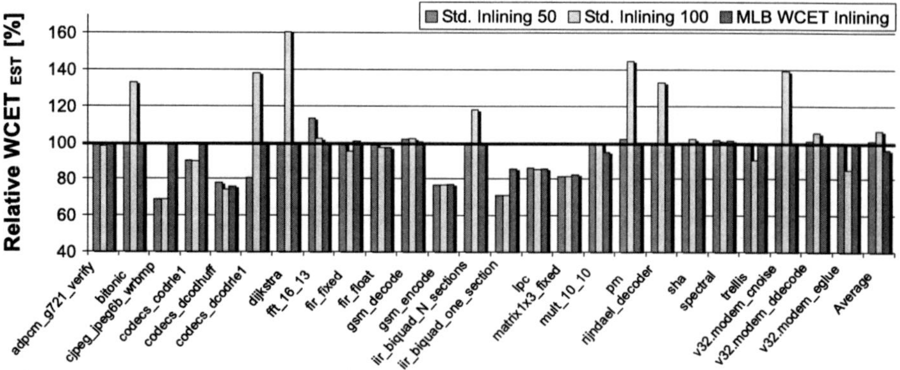

Fig. 6.5 Relative WCET estimates for WCET-aware inlining (test set)

inlining is prevented. In contrast, the standard heuristic performs inlining of function *icrc1*, leading to additional spill code within a loop that originally contained the call site. The benchmark *expint* exhibits a strong WCET reduction of 69.0% w.r.t. standard inlining. The reason for this WCET improvement is that the application of the MLB heuristics enables further optimizations, such as *constant propagation* and *constant folding*, that significantly improve the code quality. The only negative examples encountered during the experiments arose for the benchmarks *matmult* and *prime* where the MLB heuristic prevented inlining although it turned out to be beneficial as can be seen for standard inlining. This inaccurate decision can be attributed to the small number of feature vectors (61 positive examples) used to train the model for classifying an inlining candidate as beneficial. On average, WCC's WCET-aware function inlining outperformed standard inlining with the 50- and 100-expression restriction by 9.6% and 17.7%, respectively.

To provide a more accurate assessment of the generalization ability of the trained model, the *holdout* method was utilized. In this method, the training set is used to derive the model, and the validation is performed on *unseen* benchmarks from the test set. A leave-one-out cross validation was not performed since this validation type requires a tight integration of a machine learning tool into the compiler. However, this integration was avoided on purpose in this scenario where the application of the generated MLB heuristics should be independent of a machine learning tool.

Figure 6.5 shows the results of the holdout method on the WCET estimation. Due to space constraints, a subset of 25 representative benchmarks is depicted. As in the previous figure, the 100% mark represents the estimated WCET of the benchmarks compiled with *O3* and disabled function inlining. The *Average*-bar represents the WCET results averaged for all 40 benchmarks from the test set. The impact of the standard inlining with the 50-expression restriction yields a slight WCET decrease of 0.8% on average. Applying the more aggressive standard inlining (labeled with *Std. Inlining 100*) results again in a stronger degradation of the worst-case case performance by 7.3%.

Exploiting the MLB WCET-aware inlining heuristics for the test set leads to similar promising results as for the training set. On average, a WCET reduction of 4.0%

Table 6.3 Relative code size after inlining

	Code Size - Flash
Std. Inlining 50	101.3%
Std. Inlining 100	104.5%
MLB WCET Inlining	98.8%

was achieved. It can be seen that significant WCET degradations as experienced by standard inlining for example for benchmarks *bitonic* or *dijkstra* could be avoided. Moreover, a comparison between the results for the WCET-aware inlining achieved for the training and test set shows that the WCET reduction is smaller for the test set. This behavior was expected as the prediction is carried out for unseen data. However, in general it can be concluded that the classifier works well for new data since it outperforms the standard inlining heuristic by up to 10.7% on average. Thus, the MLB heuristic outperforms the commonly used inlining heuristics and similar promising results are likely for its future application.

6.4.4.4 ACET

To compare the impact of the machine learning based inlining heuristic on the ACET and WCET, the execution time of the generated code was measured using a cycle-true simulator. In general, the results for the simulated time turned out to be worse than for the WCET, with an average ACET reduction of 0.4% over the test set compared against the ACET for *O3* and disabled inlining. For some benchmarks, such as *lpc*, the MLB heuristic achieved a WCET reduction of 14.4%, but only an ACET reduction of 8.1%. This difference emphasizes that the new inlining heuristic is tuned towards an effective WCET reduction and may be less suitable as an ACET optimization.

6.4.4.5 Code Size

Table 6.3 shows the influence of function inlining on the code size of the benchmarks from the test set, with 100% being the code size of the benchmark compiled with *O3* and disabled inlining. Unlike standard inlining, WCC's MLB optimization could reduce the code size on average by 1.2%, while standard inlining increased the code size by 1.3% and 4.5% for the 50- and 100 expression heuristic, respectively. The reasons are twofold:

- MLB heuristic performed in total less function inlining than the standard inliner since numerous potentially dangerous situations, leading to a WCET degradation were prevented. This explains why a code size reduction was achieved compared to standard inlining.

- Some of the inlined functions are so called *one-call functions*. These are functions that are invoked exactly once within the source code. If it can be made sure that these functions are not executed externally from other tasks, as was assumed for the present experiments, then these functions can be inlined and the original function code can be removed from the source code. Afterwards, further optimizations can be applied, leading to smaller code compared to the non-inlined code.

Considering the positive effects of the MLB WCET-aware function inlining on the code size, adverse effects on the I-cache performance due to code size increases resulting from inlining can be excluded.

From this information about the code size, it can also be inferred that the major effects on the WCET, as depicted in Fig. 6.5, were not the result of a changed I-cache performance due to code expansion. Rather, different inlining behavior enabled different optimizations and led to a modified spill code generation.

6.4.4.6 Compilation Run Time

WCC's WCET-aware function inlining increases the compilation time compared to standard inlining. This is mainly due to the feature extraction. If it was predicted in the previous optimization step to inline a function, the optimized program has to be passed to aiT in order to obtain updated WCET information. In contrast, the evaluation of the MLB heuristics, which are simple `if-then-else` statements, is negligible. For the optimization of all benchmarks from the test set, an increase of the compilation time of 248% was observed. This increase is fully acceptable for the optimization of embedded systems.

The techniques presented in this section led to the publication [LGMM09].

6.5 Loop-Invariant Code Motion

In this section, the study of the potential of machine learning techniques for WCET-aware compilation is deepened. To emphasize the generality of these techniques, a different compiler optimization, namely *loop-invariant code motion* (*LICM*), is considered.

Loop-invariant code motion, also known as *hoisting*, is a popular ACET optimization that can nowadays be found in most compilers. The transformation moves computations, which produce the same results in each loop iteration, outside of the loop. Due to the reduced execution counts of the hoisted instruction, but also due to possible positive effects on the spilling behavior, LICM often provides a speedup. However, the transformation may also have an opposed impact on the spilling behavior, requiring the insertion of additional spill code. Hence, such conflicting interactions make the generation of suitable LICM heuristics difficult.

To tackle this problem, supervised learning is exploited for the generation of MLB WCET-aware heuristics for the optimization loop-invariant code motion. The

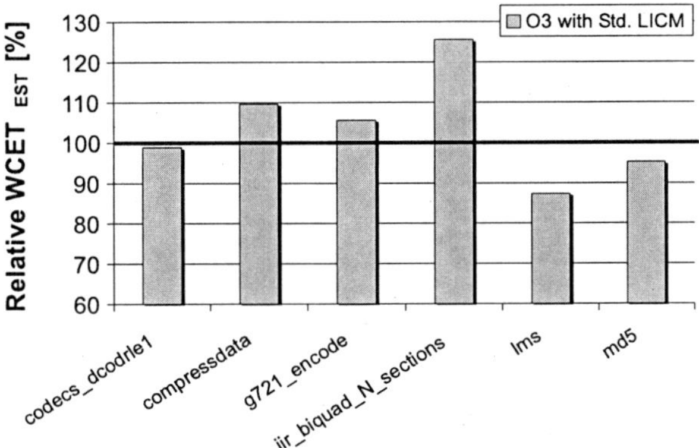

Fig. 6.6 Impact of loop-invariant code motion on WCET

main differences between the work presented in this section and Sect. 6.4.3 are the following:

- Unlike function inlining, LICM is performed in WCC at assembly level. Applying machine learning techniques also at this abstraction level of the code demonstrates that these techniques are generic and can be utilized at any level in the compiler.
- While the MLB inlining heuristics focused on understandability and an easy integration of the heuristics into the compiler, methods proposed in this section aim at maximal performance of the induced models.
- To find the best model, various learning algorithms and their parameter settings are systematically evaluated. This approach makes a seamless integration of a machine learning tool into the WCC framework necessary.
- Since accuracy may not be the most appropriate performance measure of the induced models in some situations, the evaluation of prediction models for LICM is based on the estimated WCET.

The remainder of this section is organized as follows. In Sect. 6.5.1, the application of LICM is motivated by the impact of the transformation on real-life benchmarks. Section 6.5.2 introduces the standard loop-invariant code motion in detail and highlights the optimization's positive and negative effects on the program code. Section 6.5.3 discusses the respective heuristic generation at assembly level in more detail. Furthermore, solutions to the model selection problem, i.e., which learning algorithms and parameters to select for the model induction, are presented. Finally, experimental results to evaluate the novel MLB WCET-aware LICM are provided in Sect. 6.5.4.

6.5.1 Motivating Example

The prevalent strategy to apply LICM whenever possible may degrade performance. Figure 6.6 shows the estimated WCET for six representative benchmarks compiled with *O3* and enabled standard LICM. 100% marks the WCET of the benchmarks without LICM. As can be seen, LICM may have both a positive and negative impact on the worst-case performance. For example, the benchmark *iir_biquad_N_sections* suffers from additional spill code after LICM, leading to a WCET increase of 25.5%. In contrast, the estimated WCET of benchmark *lms* decreases by 12.7% due to a reduced frequency of the loop-invariant instructions and a simultaneously positive effect of LICM on the register pressure making some spill instructions redundant.

To tackle the difficult task of finding heuristics for loop-invariant code motion, supervised machine learning is again exploited. The goal is to find a heuristic that exploits the positive effects of LICM on the one hand and prevents the transformation for adverse situations on the other hand. In contrast to related works dealing with optimizations for which different heuristics are well-studied, such as loop unrolling, this work can not benefit from any studies which describe strategies for promising LICM heuristics.

6.5.2 Standard Loop-Invariant Code Motion

Loop-invariant code motion recognizes computations within a loop that produce the same result each time the loop is executed. These computations are called *loop-invariant code* and can be moved outside the loop body without changing the program semantics [Muc97, BGS94]. If a computation occurs inside a nested loop, it may produce the same result for every iteration of the inner loop for each particular iteration of the outer loop, but different results for different iterations of the outer loop. The loop invariance of an instruction is defined as follows [App97]:

Definition 6.3 (Loop-invariant) An instruction i is said to be *loop-invariant* iff for each operand o of i: (a) o is constant, or (b) all definitions of o that reach i are located outside the loop, or (c) there is exactly one definition of o that reaches i and that definition is loop-invariant.

The detection of loop-invariant computations is straightforward if standard compiler analyses are available. Besides control-flow analysis, *def-use-chains* [ASU86], which are a common representation of data flow information about variables, allow to determine variable definitions that affect a given use of a variable.

LICM can be applied at source code level to expressions, or at assembly level, in particular to addressing computations that access elements of arrays. For an easier understanding, an example for the application of loop-invariant code motion at source code level is depicted in Fig. 6.7, with loop invariants being underlined. During the transformation, the loop-invariant expressions are moved outside the inner and outer loop, leading to the code shown on the right-hand side of the figure.

```
                                                t1 = 3*1;
for ( i=1; i < 100; ++i ) {                     for ( i=1; i < 100; ++i ) {
  m = 2*i + 3*1;                                   m = 2*i + t1;
  for ( j=1; j < i-1; ++j ) {                      t2 = i - 1;
    a[i,j] = j + m+3*i;              →            t3 = m + 3*i;
  }                                                for ( j=1; j < t2; ++j ) {
}                                                    a[i,j] = j + t3;
                                                   }
                                                 }
```

Fig. 6.7 Example for loop-invariant code motion

The positive effects of LICM are as follows:

- The shifted loop invariants exhibit a reduced execution frequency.
- The transformation may shorten variable live ranges leading to a decreased register pressure. This circumstance may in turn reduce the number of required spill code instructions.
- Moving code outside a loop reduces the loop's size which may be beneficial for the I-cache behavior since more loop code can reside in the cache.

Considering the optimized code in Fig. 6.7, LICM does not only reduce the execution frequency of the expression $3*1$ but also decreases the live range of variable 1, if 1 is not used afterwards.

Besides these positive effects on the code, LICM may also degrade performance. This is mainly due to two reasons. First, the newly created variables to store the loop-invariant results outside the loop, such as t1 on the right-hand side of Fig. 6.7, may increase the register pressure in the loops since their live ranges span across the entire loop nest. As a result, possibly additional spill code is generated. This is an issue which is especially relevant for embedded systems with small register files.

The second reason for performance degradation due to LICM is that moving the loop-invariant code might lengthen other paths of the control flow graph. This situation can be observed if the invariants are moved from a less executed to a more frequently executed path, e.g., moving instructions above a loop's *zero-trip test*.

Issues like the impact of LICM on the register pressure emphasize the dilemma compiler writers are faced with during the development of good heuristics. Performing loop-invariant code motion on the expression $3*1$ has conflicting goals, and it can not be easily predicted if this transformation is beneficial.

A review of standard compiler literature [ASU86, Muc97, CT04] revealed that LICM heuristics are unknown or at least not published. Also, there is no related work concerning this optimization except standard algorithms describing how the optimization can be realized in practice. On the software side, a similar situation can be observed. Most compilers do not model the complex interactions between different parts of the code and the loop invariants, but perform LICM whenever invariants are found without using any heuristics that might avoid the adverse effects.

6.5.3 MLB Heuristic Generation at Assembly Level

In this section, the generation of MLB heuristics for the optimization WCET-aware loop-invariant code motion applied at assembly level is discussed. Like for function inlining presented in the previous section, the heuristic generation conforms with the workflow depicted in Fig. 6.1 on p. 164. One of the key aspects of this workflow is the selection and configuration of a machine learning algorithm for the induction of the binary classification model. In general, the model selection problem can be formulated as follows:

Problem 6.1 (Model selection) Given a set of parametric machine learning algorithms $\mathcal{ML}$ and a set of parameter vectors $\mathcal{PAR}$, with $\forall alg \in \mathcal{ML} : \exists P_{alg} \in \mathcal{PAR} : P_{\text{alg}} = (p_1^{\text{alg}}, \ldots, p_m^{alg})$ and $m = |P_{alg}|$. All parameters p_i^{alg} can be assigned a parameter value $v \in Val$. For an algorithm $alg \in \mathcal{ML}$, a parameter configuration $C_{P_a} \in P_a \to Val^m$ is a function $C_{P_a} = (p_1^{alg} \to v_1, \ldots, p_m^{alg} \to v_m)$. The model selection problem for a given training set E is to find $a_{opt} \in \mathcal{ML}$ and $C_{opt} \in P_a \to Val^m$ such that the given performance function *EVAL* is optimized.

For WCC's MLB function inlining, which represented the first feasibility study of machine learning techniques for WCET reduction, the model selection problem was not explicitly considered. The selection of the learning algorithm random forests was mainly motivated by the requirements of an easy integration into the compiler and understandability of the model. A model selection driven by such factors is up to now common practice in the compiler community.

However, there is no guarantee that the employed learning algorithm yields good performance in terms of WCET reduction. The problem is the complex structure of learning algorithms and the non-trivial impact of their parameters on the program performance when used for prediction. As a consequence, the performance of the induced model can not be predicted statically with sufficient precision. Also, there is no *golden rule* for determination which learner will maximize a given performance function [BGCSV09, Bis08]. The only reliable way for a performance assessment is to generate a heuristic and to evaluate it on a set of benchmarks [LBO09].

Hence, the current state for the MLB heuristic generation can be seen as a *trial-and-error* approach. The compiler writer chooses a learner, induces a prediction model and evaluates the impact of the generated heuristic on benchmarks. If the heuristic did not yield the expected performance results, the compiler writer either tunes the learner parameters or even selects another learner and repeats the evaluation hoping for better results. Obviously, repeatedly evaluating different learners manually is time-consuming, error-prone, and it is often not clear if further tuning pays off. In literature [MBQ02, SA05, MM99, CM04, LBO09], typically one or two learners are employed without a detailed reasoning why exactly these algorithms including their parameters were chosen.

The exploitation of machine learning for heuristic generation is attractive since it helps to reduce the effort of compiler development. However, the effort is now shifted from the manual tuning of heuristics to the model selection problem of learning. To address this well-known shortcoming [SS07], an automatic model selection is proposed in this book.

6.5.3.1 Automatic Model Selection

In this section, the model selection problem is tackled in a systematic way. The current framework depicted in Fig. 6.1 on p. 164 is extended by an automatic evaluation of the considered learning algorithms and their parameter settings. To this end, a tight integration of a machine learning tool into the WCC framework is mandatory.

Learning Algorithms   The first step of the presented approach is the choice of algorithms to be involved in the performance assessment. For the present study, popular learning algorithms are included that have been successfully applied in the past for various studies and that differ in their functionality, i.e., they train a classifier based on different principles. Moreover, the learners' parameters have to be taken into account since they significantly contribute to the performance of a model as will be shown later. The following algorithms and their respective parameters [Bis08] are involved:

1. **Decision Trees** partition the examples into axes-parallel rectangles by recursively splitting the training set examples into sub-trees (for more details, see Sect. 6.4.3.3). Frequently, the gini index is used as splitting criterion. Additionally, there may be stopping criteria like the maximal depth of a tree and thresholds for the minimal number of examples in a node for its further split, the minimal number of examples in a leaf, the minimal information gain for splitting, and the number of alternative nodes considered (prepruning alternatives). Furthermore, a confidence level for post tree pruning can be specified.
2. **Random Forests** consist of several unpruned decision trees which are constructed from different bootstrap samples (for more details, see Sect. 6.4.3.3). The algorithm provides two parameters: the number of trees in the forest and the number of considered features for node splitting.
3. **Linear Support Vector Machines (SVM)** find a hyperplane which separates the examples such that those with the label $y = +1$ are in the positive half and those with the label $y = -1$ are in the negative half of the instance space. The hyperplane is determined by $W \cdot X + \beta_0$. The learning task is to estimate the weight vector W and β_0, such that the error is minimal (i.e., the instances represented by the feature vector X are placed on the correct side of the hyperplane) and that the distance between the closest instance and the hyperplane is maximal. Those examples which are closest to the hyperplane are called support vectors. In order to allow some misclassified instances, SVM offers a soft margin parameter C which gives a weight to the error. Internally, SVM compares all examples pairwise using a kernel function. For the linear SVM, the kernel function is the dot product $W \cdot X$.

4. **SVMs with Radial Base Function (RBF) Kernel** operate on not linearly separable data by including another kernel function into the learning problem of SVM. RBF covers areas of instances by a Gaussian distribution: $K_{RBF}(x_i, x_j) = \exp(\gamma(x_i, x_j)^2)$. The parameter of the Gaussian's width, γ, is decisive: for a low γ, almost every example is covered by its own RBF region, for a large γ, interesting regions cover a set of examples.

5. **k-Nearest Neighbor (kNN)** stores all examples and classifies a new input by looking at k most similar examples. The majority class of these k examples becomes the predicted class. If k is too small, the error is reduced, but the prediction becomes biased, e.g., by outliers. If k is too large, the error might also become large. Thus, the setting of the appropriate k is crucial for the learner's performance.

6. **Naive Bayes** predicts for a feature vector X the class y_i such that the likelihood $P(y_i|X)$ is maximal. According to Bayes' theorem, it is sufficient to maximize the probability $P(X|y_i)P(y_i)$, since the a priori probability of the labels in Y (e.g., $P(y_i = YES)$ or $P(y_i = NO)$) are the same for all training examples. Implicitly, Naive Bayes assumes the independence of all example's features. Due to its simple calculation, Naive Bayes is a very fast algorithm and has typically no parameters for configuration.

Performance Evaluation   There are different metrics for performance evaluation of learners. The standard performance measure of learning algorithms is accuracy (cf. Sect. 6.3.1). It is calculated on the basis of the examples generated for the test set by comparing predicted and true labels.

In case of compiler optimizations, program run time is crucial. For embedded systems acting as real-time systems, the main goal is to find a learner that yields highest WCET reduction. The performance evaluation of a learner based on the accuracy may not be appropriate since it does not allow to draw conclusions about the program's WCET. This circumstance is shown in the following example.

Example 6.1 It should be assumed that during optimization a heuristic has to decide whether the optimization steps A, B, C, and D are performed. The impact of these optimizations on the (worst-case) execution time is shown in the following table:

Optimization Step	Performed/YES	Not Performed/NO	Correct Prediction
A	−10 cycles	0 cycles	YES
B	−10 cycles	0 cycles	YES
C	−20 cycles	0 cycles	YES
D	50 cycles	0 cycles	NO

For example, if the optimization step A is performed, which corresponds to the classification prediction *YES*, then the execution time is reduced by 10 cycles. Otherwise, if the prediction for A is *NO*, then the program is not modified and the esti-

---

Algorithm 6.1 Algorithm for benchmark-wise cross validation

Input: *Learning algorithms $\mathcal{ML}$, Training set E*
Output: *best alg, s*
 1: $WCET_{ref} \leftarrow computeWCET_{ref}(E)$
 2: **for all** *algorithm alg $\in \mathcal{ML}$* **do**
 3:    **for all** *parameter setting $s \in \mathcal{PAR}$* **do**
 4:     *performance $\leftarrow 0$*
 5:     **for all** *benchmark $b \in E$* **do**
 6:      *generateHeuristic$_{MLB}$(alg, s, E \ b)*
 7:      $WCET^b_{MLB} \leftarrow computeWCET_{MLB}(b)$
 8:      *performance $\leftarrow$ performance $+ \dfrac{WCET^b_{MLB}}{WCET^b_{ref}}$*
 9:     **end for**
10:     *evaluation[alg][s] $\leftarrow \dfrac{performance}{|E|}$*
11:    **end for**
12: **end for**
13: **return** min *(evaluation)*

---

mated WCET remains unchanged. The last column of the table depicts the correct predictions leading to the best optimization result.

It should be further assumed that models M_1 and M_2 induced by two machine learning algorithms are used for the prediction of the given optimization steps. Model M_1 predicts A, B, and C correctly, leading to an accuracy of 75% and an overall increase of the program's WCET of $(-10 - 10 - 20 + 50 =)$ 10 cycles. In contrast, model M_2 predicts C and D correctly, yielding a smaller accuracy of 50% and an overall positive impact on the WCET of $(0 + 0 - 20 + 0 =)$ -20 cycles.

Thus, based on accuracy as performance function the wrong conclusion of selecting M_1 as the more promising model would be drawn.

Due to this missing correlation between the accuracy and the program (worst-case) execution time, learners should be evaluated by directly determining the program performance instead of their accuracy.

Moreover, the classical N-fold cross validation has to be applied in a modified fashion for the performance evaluation of learners used as optimization heuristics. An adequate performance evaluation for the model selection problem (cf. Problem 6.1) is the so-called *benchmark-wise cross validation*. Its structure is presented in Algorithm 6.1.

In the beginning, for each benchmark of the training set E, a reference WCET, such as the WCET of b when the considered optimization is completely disabled, is determined (line 1). Next, for each considered learning algorithm $alg \in \mathcal{ML}$ (line 2) and its parameter setting $s \in \mathcal{PAR}$ (line 3), a per-benchmark validation is conducted. To this end, for each learner/parameter combination a model is induced using all benchmarks from the training set except benchmark b (line 6). To determine the impact of the generated MLB heuristic, the estimated WCET using the MLB heuristic is compared against the respective reference WCET (lines 7–8).

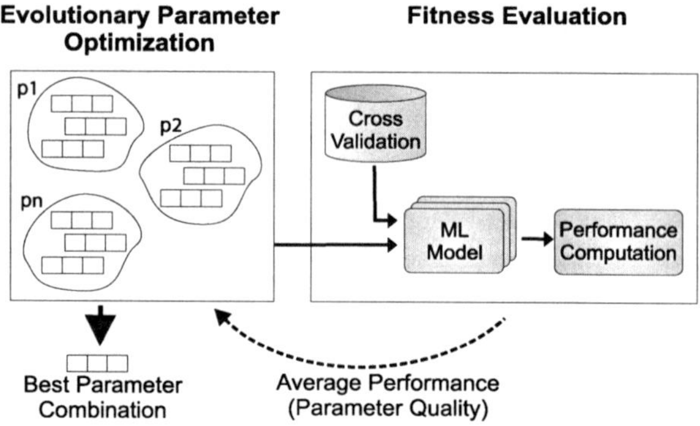

Fig. 6.8 Evolutionary parameter optimization

This cross validation is performed for all benchmarks of the training set, and the determined average value represents the impact of an MLB heuristic on the WCET using a particular learner *alg* and the respective parameter settings *s* (line 10). Finally, Algorithm 6.1 returns *alg* and *s* with the minimal relative WCET (line 13), i.e., the maximal WCET reduction. Hence, the computation in lines 4–10 represents the performance function *EVAL* of Problem 6.1 that is to be minimized.

Using this *benchmark-wise* cross validation is a common approach to estimate the *generalization* ability of a learner, i.e., by applying the models to *unseen* benchmarks, it can be inferred how well new examples will perform using this model.

Parameter Optimization The naive approach shown in Algorithm 6.1, which iterates over all parameter settings $\mathcal{PAR}$, can not be feasible in practice since the number of user-defineable classifier parameters is too large to be exhaustively searched. To tackle this problem, an evolutionary strategy [Mie08] is proposed that substitutes the complete set $\mathcal{PAR}$ in Algorithm 6.1 by a subset of promising parameter settings.

This approach is visualized in Fig. 6.8. Each individual in a population p_n of size (*pop_size*) represents one specific parameter value for a particular learning algorithm, such as value for C and γ in case of the SVM with RBF kernel.

In the beginning, for each learning algorithm, populations of randomly initialized individuals are generated. Next, each individual is passed to the benchmark-wise cross validation. The averaged performance value serves as the fitness function for the individual under analysis.

In the subsequent crossover step, individuals mate with a specified probability (*crossover_prob*) using a single cross-over point on both individuals. The produced children contain the exchanged parameter values of their parents. These children are added to the current population. Then, all individuals are cloned and the clones are mutated by adding values from a Gaussian distribution to all parameters.

The newly created individuals during crossover and mutation are passed to the benchmark-wise cross validation to determine their fitness. To construct populations for the new generation, individuals from the previous generation as well as the new individuals take part in the tournament selection. The selection chooses individuals with highest performance (as parents) as long as the number of selected individuals does not exceed *pop_size*.

The parameter optimization maintains the best individuals (*elitist selection*) and terminates if either a specified maximum number of generations (*max_gen*) is reached or there was no improvement over *imp* generations. Hence, the underlying genetic algorithm of WCC's parameter optimization is an adoption of standard genetic algorithms [Hol92].

6.5.3.2 Feature Extraction

The WCC framework for the automatic selection of machine learning models is generic, i.e., it can be exploited to generate heuristics for a large number of low-level optimizations without any major adaption. To enable this option, a large set of features extracted from the compiler must be provided. These features must be chosen such that they cover a wide range of various characteristics of the program.

For the generation of MLB heuristics for loop-invariant code motion, the features characterize each instruction being an LICM candidate as well as the basic block to which the instruction is moved. WCC's feature extractor generates 73 features in total which describe characteristics of single instructions, basic blocks, loops, or functions, depending on which low-level construct is passed to the feature extractor. The features can be classified as follows (given some examples):

1. **Structural features**: Type of instruction (arithmetic, load/store, jumps, floating point, etc.), size of given construct, number of block successors/predecessors, number of operands in given instruction
2. **Liveness analysis related**: Liveness information (*live-in* and *live-out*) of instruction, number of *defs* and *uses* in instructions/blocks, information about register live times (for register pressure estimation)
3. **Loop features**: Loop nest levels, loop iteration counts
4. **Misc**: Length of critical path in loop, outcome of static branch prediction for jump instructions

This set of features is variable, i.e., depending on the application, all features or just a subset can be used. The feature extractor was designed in a flexible way such that new features can be easily added. For learning algorithms that can only handle numerical values, nominal features are first transformed into discrete numerical values and then normalized by a linear transformation into [0, 1].

6.5.3.3 Label Determination

After feature extraction for the loop-invariant instruction i_{inv}, the label is determined as follows. Before hoisting i_{inv}, the WCET is estimated for the embedding code region *reg* of i_{inv}. The definition of *region* depends on whether i_{inv} was initially located in a nested loop or not. If i_{inv} is hoisted from an inner to an outer loop, region *reg* is represented by the basic blocks of the outer loop. Otherwise, if i_{inv} is moved outside a non-nested loop, then *reg* consists of all blocks located in the function f, with $i_{inv} \in f$.

Using outer loops in case of nested loops for *reg* instead of the entire function makes the label extraction more reliable since it captures the effects of a particular LICM more precisely. A decreased WCET after LICM means that the transformation is beneficial (labeled with *YES*) for i_{inv} in its current context. The context is expressed by the static features of the target block b_{target} to which the instruction i_{inv} is moved. If the WCET does not change, also the label *YES* is used to perform code motion which possibly enables optimization potential for subsequent LICM candidates. If a WCET increase due to adverse LICM effects was identified, the feature vector is labeled with *NO* to indicate that the code motion should be avoided for similar cases. For the next LICM candidate, i_{inv} is kept in its new position and the next loop-invariant instruction is considered.

6.5.3.4 Application of WCET-Aware LICM

In contrast to WCET-aware function inlining discussed in Sect. 6.4, WCET-aware LICM employs more flexible heuristics. The heuristics are not restricted to a small number of learning algorithms that induce models which can be easily integrated into a compiler as `if-then-else` constructs. Rather, any supervised learner deriving a classifier model is allowed.

To provide this flexibility, a tight integration of a machine learning tool into the compiler framework is mandatory. In the beginning of loop-invariant code motion, the machine learning tool is initialized with the best model determined during WCC's automatic model selection, and a socket connection to the WCC framework is established. Next, WCC's LLIR optimizer traverses the code in a depth-first search manner and extracts for each LICM candidate its static features. This feature vector is sent to the ML tool that returns the *YES/NO* prediction. Based on this information, the considered loop-invariant instruction is shifted or not. LICM is repeated from the beginning as long as loop-invariants are moved.

6.5.4 Experimental Results for WCET-Aware LICM

To demonstrate the practical use of WCC's WCET-aware LICM, evaluation on a large number of different benchmarks was conducted. The 39 benchmarks come from the test suites DSPstone [ZVS+94], MediaBench [LPMS97], MiBench [GRE+01], MRTC [MWRG10], NetBench [MMSH01], and UTDSP [UTD10]. On the one hand, the benchmarks were used to construct the data set for machine learning which serves as training data for the LICM heuristic generation. On the other hand, this data was used in the cross validation phase to evaluate the performance of the heuristics for WCET reduction.

The training set based on these benchmarks comprised 3,491 examples (LICM candidates), and its construction took about 50 hours on two Intel Xeon 2.13 GHz quad cores. It should be noted that this construction represents a unique overhead since the data set construction has to be performed only once off-line and can be reused as long as the internal compiler structure or the target architecture is not modified.

The evolutionary parameter optimization as part of the automatic model selection was performed for each of the six machine learners discussed in Sect. 6.5.3.1. The following parameter settings were used for the genetic algorithm (cf. p. 186):

- population size *pop_size*: 20
- number of generations *max_gen*: 5
- tournament selection performed on 30% of population size
- crossover probability *crossover_prob*: 90%
- Gaussian mutation, i.e., adding a random value from a Gaussian distribution to each element of the individual
- termination if no improvement for 2 generations was observed or *max_gen* is reached

Some of these parameter settings, such as for *crossover_prob*, are default settings for a search via a genetic algorithm, while other settings, like *pop_size*, were used to restrict the search time.

For the benchmark-wise cross validation (cf. Algorithm 6.1), the reference value $WCET_{ref}$ represents the estimated WCET of the benchmark under analysis optimized with highest optimization level (*O3*) and disabled standard LICM. In contrast, $WCET_{MLB}$ is the WCET estimation for this benchmark optimized with *O3* using LICM with the new MLB heuristic.

Most of the time for the evolutionary search was consumed by the WCET analyses. For one run of the benchmark-wise cross validation, i.e., inducing 39 models and using them for the WCET estimation of each benchmark in the test set, about 8 minutes on two Intel Xeon 2.13 GHz quad cores of a system with 8 GB RAM were required. Depending on the evolution of the evolutionary search, the maximal run time of 19 hours was observed for the evaluation of the learner random forest.

All experiments were performed in the WCC framework using TriCore's cached Flash memory. The employed compiler framework, without modules not relevant for MLB WCET-aware LICM, is depicted in Fig. 6.9. As can be seen, the feature extraction is performed in the compiler backend and provides input for the machine learning tool RapidMiner [MWK+06] that delivers the prediction for LICM.

6.5.4.1 WCET

In a first phase, the machine learning model selection was performed to find the best learner. Table 6.4 gives an overview of the considered learners, their parameters, and the explored parameter values by the evolutionary search (column *Range*). It should be noted that Naive Bayes does not provide any parameters to be optimized.

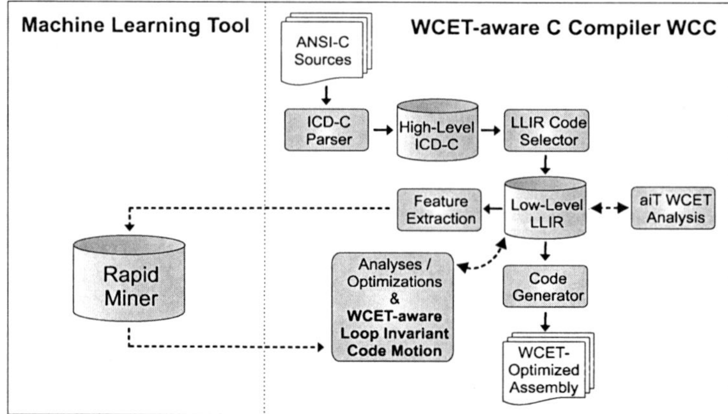

Fig. 6.9 WCC framework with integrated machine learning tool

Table 6.4 Learning algorithms with possible and best parameter settings

Parameter	Range	Best	Parameter	Range	Best
Decision Trees			**Linear SVM**		
max. depth	[1;20]	16	C	[0;10,000]	616.11
min. split size	[4;100]	19	**SVM with RBF kernel**		
min. leaf size	[2;100]	31	C	[0;10,000]	2,405.15
min. gain	[0;0.03]	0.014	γ	[0;74]	30.081
prepr. altern.	[3;10]	4	**kNN**		
confidence	[0.1;0.5]	0.476	k	[3;100]	11
Random Forests			**Naive Bayes**		
no. of trees	[1;100]	7	*no. configurable parameters*		
features	[1;73]	39			

However, the algorithm was considered due to its popularity and its specific functionality. Due to the large search space of possible parameter values, it becomes obvious that an exhaustive search is infeasible.

Table 6.5 summarizes the results of the evolutionary parameter optimizations for the six considered learners. The results in the second, third, and fourth column represent the performance values, i.e., the averaged relative WCET results obtained during the benchmark-wise cross validation when comparing the WCET using the MLB heuristic with the WCET of code compiled with *O3* and disabled LICM (corresponding to 100%).

In more detail, the second column (*Best*) represents the highest improvement of the WCET observed during the evolutionary search of each learner. These values were achieved using the parameter combinations shown in the third column of Table 6.4. For example, 95.36% for *SVM with RBF kernel* means that the WCET was reduced on average by 4.64%. The third and fourth column (*Worst, Average*) of Table 6.5 depict the worst and average WCET reduction (over all runs) found by the evolutionary search. Finally, the last column (*Accuracy*) describes the classification accuracy that was computed for the parameter combination leading to the best

timized code (*O3*) and disabled LICM. The light bars show standard LICM with the common LICM strategy that performs the code motion whenever possible. The dark bars represent the cross validation results, i.e., the MLB heuristic was trained without the respective benchmark and afterwards applied to it. By learning a model and validating it on the excluded benchmark, the dark bars indicate how good the heuristic performs on *unseen* data. As can be seen in the figure, in most cases the new MLB WCET-aware LICM outperforms the standard LICM optimization, with up to 36.98% for the *fir* benchmark from the MRTC suite. On average, the standard LICM achieves a WCET reduction of merely 0.56%, while WCC's MLB WCET-aware LICM reduces the WCET by 4.64%, as already shown in Table 6.5.

In few cases, standard LICM could slightly outperform the MLB heuristic. For example, in case of benchmark *lpc* (from the UTDSP suite) the decision taken by the novel heuristic led to a situation in the code that requires addition spill code, increasing the WCET by 5.16%. On the contrary, standard LICM hoisted all loop-invariant instructions resulting in a WCET decrease of 2.63%. Such scenarios may occur due to mispredictions of the induced model and can be never completely excluded.

6.5.4.2 ACET

Similar to the WCET results, the impact of different LICM strategies on the ACET was measured using a cycle-true simulator. For the analyzed 39 benchmark, standard LICM slightly degraded the ACET on average by 0.89%. In a similar way, the MLB heuristic yielded a marginal ACET increase by 0.48% on average. Even though the new heuristic marginally improves the standard LICM strategy, the results for the ACET reduction are poor compared to those achieved for the WCET reduction. These results allow to draw the conclusion that an effective reduction of a particular performance metric, such as the ACET or WCET, can be only accomplished if this metric is also involved in the automatic model selection. Therefore, heuristics generated for WCET reduction may not exhibit comparably good results for ACET reduction and vice versa.

6.5.4.3 Compilation run time

The communication between WCC and RapidMiner is established in an efficient way using a socket connection. However, this communication still introduces additional overhead. Furthermore, each LICM decision requires a feature extraction. As a result, the compilation run time of all benchmarks from the test set using MLB WCET-aware LICM increases by 136% compared to standard LICM, which is fully acceptable for embedded system optimization. Most of this overhead is caused by the interaction with the machine learning tool.

The techniques presented in this section have been published in [LSMM10].

6.6 Summary

With the growing complexity of software, manual optimization and tuning of programs for an improved performance became a difficult and tedious task. To relieve software developers from this burden and to allow them to concentrate on writing correct software, optimizing compilers have been developed. Today, compiler developers ironically face similar problems. The development of highly effective optimizing compilers becomes a challenging but also a tedious mission. This is due to the non-trivial interaction between code optimizations and the complex architecture of today's systems on the one hand and the mutual, sometimes conflicting, interactions between optimizations on the other hand. Therefore, efficient solutions to this dilemma are desirable, allowing compiler writers to focus again on the development of correct and effective optimizations.

In this chapter, machine learning techniques for an automatic generation of compiler heuristics were studied for the first time in the context of WCET reduction. One of the main advantages of these techniques is their simple application. Based on automated observations of the impact of compiler optimizations on the worst-case performance of representative benchmarks, appropriate heuristics can be automatically learned—even for complex systems where human understanding is limited. Another advantage of machine learning techniques is their effectiveness, often leading to heuristics that outperform their hand-crafted counterparts. Therefore, the application of these techniques does not only reduce development time but also helps to construct high-performance compilers.

The first optimization studied in this chapter was the source code optimization function inlining. Using WCC's infrastructure, static features, which characterize a particular inlining candidate, were gathered from the high-level representation and the compiler backend using WCC's back-annotation. Based on these features, the supervised learning algorithm random forests generated an inlining heuristic that was directly integrated into the source code of WCC's optimizer. Regarding the WCET reduction, the novel MLB WCET-aware heuristic could outperform standard inlining on average by 11.4% and 10.7% for the training and test set of benchmarks, respectively.

In a further study, machine learning based heuristics were generated for the optimization loop-invariant code motion. Unlike WCC's inlining, this optimization is applied at the assembly level to show that the presented techniques work well on any abstraction level of the code. Moreover, the well-known problem of model selection—i.e., selecting learning algorithms and their respective parameters—was explicitly addressed. Using an evolutionary search and systematically evaluating the performance of the learners based on their impact on the WCET, enabled the generation of effective LICM heuristics that outperformed standard LICM by a factor of up to 8.3.

The two studied standard compiler optimizations represent a vast class of optimizations with conflicting goals, i.e., code transformations that may improve but also degrade program performance. Their traditional heuristics can be often improved by machine learning. However, as the experiments in this chapter showed, an MLB heuristic trained for a particular objective, such as WCET, may exhibit poor performance for another objective. Therefore, it can be concluded that the current state of today's compiler-based WCET reduction can be further improved if additional standard ACET optimizations are extended by MLB WCET-aware heuristics.

Chapter 7
Multi-objective Optimizations

Contents

7.1 Introduction

The development of high-performance optimizing compilers is an inherently hard problem for many reasons. In the previous chapter, one of the reasons, namely insufficient compiler heuristics, was addressed. It was shown that supervised machine learning can be exploited to automatically generate efficient heuristics that are able to outperform traditional heuristics. Another key issue why compilers do not deliver optimal performance are interactions between optimizations, i.e., the effect of an optimization may depend on the presence or absence of another optimization in the compilation sequence.

To relieve the compiler user from a search for appropriate compiler optimizations, conventional compilers are equipped with standard optimization levels such

P. Lokuciejewski, P. Marwedel, *Worst-Case Execution Time Aware*
Compilation Techniques for Real-Time Systems, Embedded Systems,
DOI 10.1007/978-90-481-9929-7_7, © Springer Science+Business Media B.V. 2011

as *O3*. These levels are generated by compiler writers based on their experience and intuition—a process that may be highly challenging as shown in [ITK+03]. However, despite this enormous effort, there is no guarantee that these optimization levels perform well for other architectures and benchmarks. Almagor et al. [ACG+04] observed that the performance of standard optimization sequences in many compilers are is below that of best performance-specific optimization sequences. Even adverse effects of these sequences on the program performance were reported [ABC+06, KHW+05, ZCS03, CFA+07]. A reason for this failure is that, despite years of research, the nature of these interactions is only partially understood [SS07], thus many interactions are ignored during compiler design.

This shortcoming of compilers in the past led to a search for promising compiler optimization sequences, known in literature as the *phase ordering* problem. The idea behind phase ordering is to generate tailored optimization sequences that are adapted to the compiler's environment, that is, the capability to reconfigure the compiler in such a way that it meets the requirements of a single or a set of applications being executed on a specific target architecture. To enable this flexibility, the rigid structure of conventional compilers with its predefined optimization order has to be extended by the capability to perform optimizations in arbitrary sequences. Such compilers are referred to as *adaptive compilers*.

For modern systems, the situation becomes even worse. In addition to the challenges of finding promising optimization sequences, modern systems have to meet different requirements. In case of embedded real-time systems, not only the ACET but also WCET, code size, dependability, and temperature are crucial. However, compiler optimizations often have a conflicting effect on these objectives, that is, the improvement of the code w.r.t. one objective may degrade another one. Hence, finding a satisfying solution is an even more complex task.

This chapter provides a complementary solution for the construction of high-performance compilers. The extended WCC does not only find good optimization sequences for a single objective function, but searches for compiler optimization levels that provide a trade-off between different objectives, namely the WCET, ACET, and code size. In this work, the class of ACET optimizations was intentionally chosen since it allows to conduct a very first systematic study on the impact of standard optimizations on the worst-case behavior of a system. Moreover, due to the use of standard compiler optimizations, the presented techniques can be easily adopted to other compilers that support similar standard compiler optimizations.

In summary, the advantages of the search techniques proposed in this book are the following:

- The automatic search for good optimization sequences relieves the compiler developer/user from this tedious and complex task.
- The determined optimization sequences clearly outperform standard optimization levels.
- The presented techniques can be easily adopted to other frameworks.
- The results provide insight to developers and researchers of real-time systems. Due to the comprehensive evaluation of optimizations, this study provides for the

first time an answer to the following question: how do standard compiler optimizations affect the WCET compared to other cost functions? Thus, the current uncertainty of real-time system developers on optimizing compilers can be clarified.

The remainder of this chapter is organized as follows. In Sect. 7.2, it is demonstrated how WCC's standard optimization levels fail to deliver optimal performance, leaving room for further optimization potential. Section 7.3 provides an overview of related work that addresses the phase ordering problem. In Sect. 7.4, the general structure of an adaptive compiler is presented and extensions to WCC, which were developed in the course of this book, are discussed, turning WCC compiler into a WCET-aware adaptive compiler framework. To cope with the huge search space of compiler optimizations, Sect. 7.5 presents an automatic approach for compiler optimization sequence exploration based on evolutionary multi-objective search algorithms. Experimental results for the exploration of compiler optimization sequences are provided in Sect. 7.6. Finally, this chapter is summarized in Sect. 7.7.

7.2 Motivation

It is well known that compiler optimizations interact, i.e., the impact of a given optimization on the program performance depends on whether other optimizations were performed as well or not. Moreover, the nature of interactions also relies on the order in which the optimizations are carried out.

Interactions can be classified into those with a positive and a negative influence on the program performance. Positive interactions typically arise when one optimization enables additional potential for other optimizations. Examples of this class are loop unrolling or function inlining as was discussed in more detail in Sects. 4.5 and 6.4, respectively.

In contrast, negative interactions occur when one optimization transforms the code such that a following optimization can not be applied anymore. A typical example is the interaction between the optimization register allocation and instruction scheduling (cf. Sect. 5.4). The application of register allocation on the code creates *false* dependencies since independent temporary variables have to be mapped to the same physical register. As a consequence, the following instruction scheduling is restricted, avoiding to utilize full parallelism to generate compact code.

But not only interactions between optimizations can degrade performance. Code transformations may modify program code such that its execution is slowed down on a particular hardware. Prominent examples are code expanding optimizations that yield I-cache overflows. For single optimizations, such as WCC's WCET-aware loop unrolling (cf. Sect. 4.5), a heavy exploitation of the entire compiler framework may provide sufficient information to avoid negative interactions with the memory hierarchies. However, in general many compilers lack a detailed description of the underlying hardware and the development of sophisticated compiler heuristics is often non-trivial as shown in Chap. 6. Thus, few interactions with the hardware are taken into account.

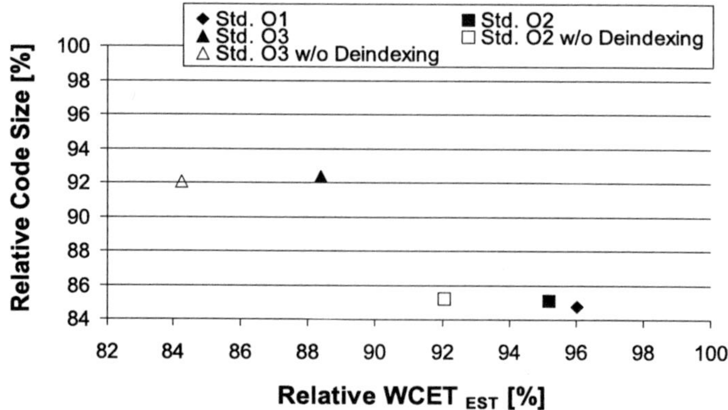

Fig. 7.1 Impact of loop deindexing on standard optimization levels

To show how standard optimization levels affect different objective functions, Fig. 7.1 depicts the impact of WCC's optimization levels on the estimated WCET and the code size. The baseline of 100% is the average performance of WCC's 35 representative benchmarks on the corresponding objective functions when no code optimizations are conducted.

The filled symbols represent WCC's standard optimization levels *O1*, *O2*, and *O3*. As can be seen, the two objective functions are conflicting. Increasing the optimization level, that is, performing more optimizations, yields a decrease of the estimated WCET but simultaneously also a code size increase. This behavior is obvious since *O3* includes code expanding optimizations, such as loop unrolling or function inlining.

In a further experiment, the source code optimization loop deindexing (cf. Sect. 4.1 on p. 62), which is included in WCC's levels *O2* and *O3*, was completely disabled. The performance results are represented in Fig. 7.1 by the symbols labeled with *Std. O2 w/o Deindexing* and *Std. O3 w/o Deindexing*. Surprisingly, the absence of this transformation has a positive effect on the estimated WCET compared to standard *O2*, while the code size remains almost unchanged. Considering the results for *O3*, disabled deindexing has even an overall positive effect on both objective functions.

The question why loop deindexing degrades overall performance in the higher optimization levels is not easy to answer. This is due to the reason that deindexing is applied in concert with other optimizations, thus complex interactions must be considered in their entirety and all factors that can influence the nature of the interactions have to be taken into account. However, due to the complexity of this problem, a human understanding is usually impossible or may lead to wrong conclusions. As a result, compiler optimization levels may fail to deliver good performance.

In summary, the results presented in Fig. 7.1 allow to draw two main conclusions:

- The standard optimization levels can be improved. A simple deactivation of loop deindexing already improved the considered objectives. Thus, using arbitrary optimization sequences, where multiple optimizations are switched off or are applied in a different order, promises space for further improvements.

- The results indicate that the estimated WCET and code size are conflicting goals. Therefore, it can be expected that other optimization sequences can further exhaust this conflict and yield significant improvements of one objective at the cost of the other. Such trade-offs could be offered to system designers who select the sequence that best suits their system.

The severity of the phase ordering problem and the additional optimization potential, which can be expected from more flexible optimization sequences, suggest a redesign of traditional compilers. The required flexibility to adapt to a given environment is offered by adaptive compilers. The following sections discuss this topic and present extensions to turn WCC into a WCET-aware adaptive compiler.

7.3 Related Work

The phase ordering problem is one of the well-known problems in the compiler domain. There is a small number of works that attempts to analytically anticipate the impact of compiler optimizations to find good phase orderings. Whitfield and Soffa [WS90] developed a framework for examining the interactions of the transformations based on an axiomatic specification technique. Using these specifications, it can be inferred which optimizations enable or cancel the application of other optimizations. Bashford [BL99] applied constraint logic programming to model a coupling of different code generation phases, namely code selection, register allocation, and instruction scheduling. In [Fur05], Fursin used efficient heuristics to search for promising optimizations in a unified transformation framework which represents a set of transformations in a unified way. Zhao et al. [ZCS03] presented a framework that aims at statically predicting the impact of applied optimizations.

Such analytical approaches allow to find optimal optimization solutions. However, their practical application is constrained by two issues. First, these approaches only allow the consideration of a selected set of optimizations for which a formal specification must be given. Second, due to complexity reasons their application may become infeasible for larger applications.

An alternative approach to search for good compiler optimization sequences is *iterative compilation*. In contrast to the analytical approach, this empirical technique does not rely on any formal models but evaluates different phase orderings by simulating the program or executing it on the hardware. Hence, all aspects of the hardware are taken into account.

The general idea behind iterative compilation is to explore the compiler optimization space by starting with a set of randomly chosen optimization sequences used to generate a binary executable. Typically, random sequences are used since good sequences as starting point are often not known. Measuring a single objective function, e.g., the ACET [ABC+06, KHW+05, ACG+04, LOW09, ZCS03, CFA+07] or code size [CSS99], the fitness of each sequence is determined and subsequent generations of optimization sequences yielding a higher fitness are computed using various search algorithms. To reduce the cost of iterative compilation resulting from the search in the large space, different solutions have been proposed. Kulkarni [KHW+05] used genetic algorithms to avoid an exhaustive search. In [ACG+04], a characterization of the search space was proposed to find good compilation sequences more efficiently. To accelerate the search, Leather [LOW09] applied fixed sampling plans. Agakov et al. [ABC+06] reduced the number of evaluations using machine learning approaches by focusing on promising areas of the search space.

All these aforementioned publications consider a single objective function. This approach is, however, not sufficient for modern embedded systems where a trade-off between different, conflicting optimization objectives is required. The only work addressing multi-objective compiler optimization level exploration was presented by Hoste [HE08]. However, Hoste's work differs in several ways from the adaptive WCC framework presented in this book. Most importantly, WCC's main focus is the improvement of the worst-case behavior of real-time systems, thus the trade-off between the WCET and other crucial objective functions (ACET, code size) is evaluated. Moreover, WCC does not rely on the performance of a single evolutionary algorithm, but evaluates different multi-objective algorithms by a statistical performance assessment to find the algorithm that performs best for a multi-objective exploration of compiler optimizations. In addition, the adaptive WCC framework is more flexible. In contrast to the GCC compiler used by Hoste [HE08], where optimizations are performed in a fixed order and can only be switched on or off, the adaptive WCC allows the construction of arbitrary optimization sequences. This leads to an increased complexity of the problem but also allows the exploitation of a higher optimization potential.

In the domain of WCET minimization, the only work addressing iterative compilation was presented in [ZKW+04]. The authors apply a genetic algorithm to find a sequence of standard assembly level optimizations that yield the highest WCET minimization. However, in contrast to the present work, the authors focus on a single objective function to be optimized and do not consider trade-offs with other objectives. Moreover, just a single evolutionary algorithm is applied. Thus, it is not clear how good the algorithm performs and whether other algorithms, like *Hill climbing*, might even produce better results. Finally, exclusively assembly level optimizations are considered, neglecting the potential of source code optimizations on the program's WCET.

7.4 Compiler Optimization Sequence Exploration

The structure of modern compilers is more or less rigid since optimizations are applied to an intermediate representation in a predefined order. This inflexibility is one of the major reasons why optimizing compilers do not deliver optimal performance. To address this problem, compilers must be extended in such a way that they can adapt to the compiled program and hardware. This adaptability is achieved

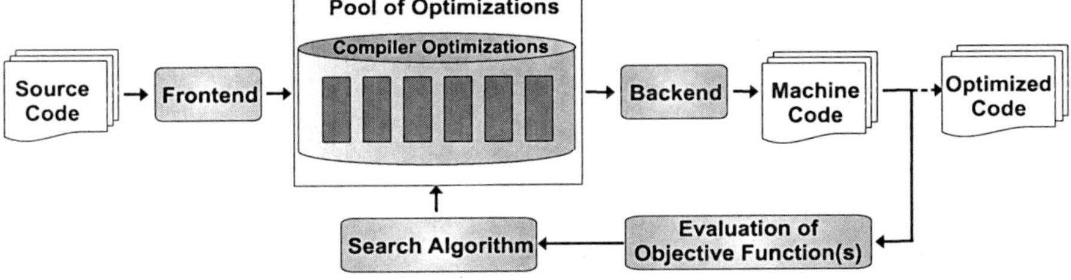

Fig. 7.2 Workflow of iterative compilation using an adaptive compiler

by searching for promising optimization sequences that lead to a maximal improvement of the considered objectives. Compilers that provide this capability are known as adaptive compilers.

This section discusses issues related to the exploration of the compiler optimization sequence search space. In Sect. 7.4.1, the general structure of adaptive compilers is introduced. Section 7.4.2 provides an overview of extensions, turning the WCC framework into an adaptive WCET-aware compiler. To find good optimization sequences, the adaptive WCC and the evolutionary multi-objective search algorithms communicate in a bidirectional way. The search algorithm generates individuals, which represent optimization sequences to be evaluated, and passes them to WCC. The compiler determines the fitness of different optimization sequences w.r.t. the considered objectives and passes these fitness values back to the search algorithm. Optimization sequence encoding and objective evaluation are presented in Sects. 7.4.3 and 7.4.4, respectively. Based on this information, evolutionary multi-objective algorithms, which will be discussed in the next section, select promising sequences, thus refine the sequence performance.

7.4.1 Adaptive Compilers

The workflow of an adaptive compiler differs from a conventional compiler in several respects. The general workflow of an adaptive compiler, as was for example proposed by Cooper [CST02], is depicted in Fig. 7.2.

Similar to standard compilers, the source code is processed by a compiler front-end that usually translates the input code into an intermediate representation, enabling an easier application of optimizations. However, in contrast to standard compilers, the optimizations are not performed in a fixed order. The search algorithm selects optimization sequences (of arbitrary order) and provides them to the compiler which generates machine code using the backend.

Next, the code is evaluated and one or more objective functions are determined, depending on whether a single- or multi-objective optimization is applied. Subsequently, the determined objective functions serve as input for the search algorithm that refines its selection of optimization sequences by choosing those optimizations for the next generation that exhibit an improved performance on the considered

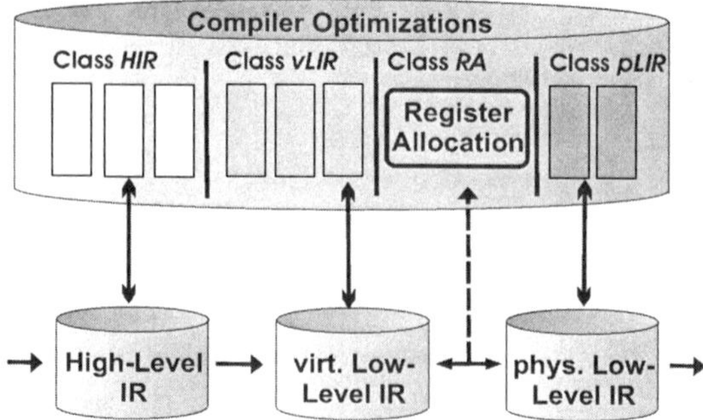

Fig. 7.3 Internal code representation and optimization within WCC

objective(s). This process is repeated until a predefined termination condition is satisfied. Finally, the best optimization sequence is applied to generate the optimized code.

Due to the enormous number of supported optimizations within modern compilers, the main problem with iterative compilation is the large search space. For example, the GNU Compiler Collection (*GCC*) v4.1 [GCC10] provides 60 compiler flags, which can be arbitrarily enabled or disabled, yielding 2^{60} possible combinations— a number of combinations that makes an exhaustive evaluation infeasible.

7.4.2 Adaptability in WCC

To provide adaptability within WCC, the compiler framework has to be extended. The general structure of the compiler, as shown in Fig. 3.1 on p. 27, remains unchanged. The extensions concern WCC's optimization modules. Therefore, to understand the concepts of iterative compilation in WCC, additional details about the internal structure of WCC's optimizer have to be provided.

7.4.2.1 Internal Structure of WCC's Optimizer

Internally, WCC manages the program code by two different intermediate representations, the high level IR *ICD-C* and the low-level IR *LLIR*. The latter can be further subdivided into a virtual and physical LLIR. *Virtual* means that no physical registers but place holders identifying dependencies among instructions are used. The translation of a virtual LLIR into a physical LLIR is performed by the register allocation that assigns each virtual register a physical CPU register.

This compiler structure yields high optimization potential since analyses and optimizations can be performed on different abstraction levels of the code. Consequently, the available WCC optimizations are subdivided into *classes*: each optimization belonging to a particular IR class $Class_{IR}$ can only be applied to the representation *IR*. This classification is depicted in Fig. 7.3. It should be noted that the

Fig. 7.4 Rigid optimizer structure of conventional compilers

```
if( enabled( optimization_A ) )
  perform( optimization_A );
if( enabled( optimization_B ) )
  perform( optimization_B );
if( enabled( optimization_C ) )
  perform( optimization_C );
...
```

register allocation is also considered as an optimization. However, due to its special role, as will be described later, it belongs to a separate IR class.

In its initial state as a conventional compiler, WCC performed the optimizations of each IR class in a rigid order. The compiler optimizer structure for each class resembles a set of consecutive *if-then*-statements as shown in Fig. 7.4.

This structure is traversed in a top-down manner. Enabled optimizations are performed always in the same sequence independent of the order they were handed to the compiler. In addition, optimizations can not be performed multiple times, such as in {optimization_A, optimization_B, optimization_A}.

To avoid a fixed phase ordering, WCC was extended such that the order of arbitrary sequences passed to the compiler is internally preserved, i.e., the optimizations are executed in exactly the same order as was used for the invocation of the compiler. The processing order of the available IR classes is still preserved, since this order is regulated by the compiler workflow. Using this flexibility in each optimization class allows WCC to adapt to optimization sequences that best suit given requirements.

7.4.2.2 Available Compiler Optimizations

In the following, WCC's optimizations considered for the compiler optimization sequence exploration are discussed in more detail.

The optimizations available within WCC are both standard ACET optimizations and WCET-aware optimizations. This study exclusively focuses on WCC's ACET optimizations for two reasons. First, the impact of standard compiler optimizations on the program's worst-case performance should be explored to show which trade-offs w.r.t. other objectives can be achieved. Using standard optimizations, the results of this study are more general and allow to draw conclusions for similar (standard) compiler frameworks. Second, WCET-aware optimizations are typically too time-consuming since they perform a costly static WCET estimation multiple times to keep their worst-case timing model up-to-date. This makes them less suitable for an iterative compilation.

WCC provides 21 standard source code optimizations that are applied to the high-level IR, hence belonging to the class *HIR* (cf. Fig. 7.3). A detailed overview of WCC's source code optimizations is provided in Table 3.1 on p. 29. In addition, some of the optimizations are parametric. For example, function inlining allows the specification of the maximal size of the callee function to be inlined. For the sequence exploration, each optimization used with a different parameter is considered as a distinct optimization. Hence, in total, 30 source code optimizations are distinguished: 18 non-parametric and 12 parametric (consisting of 3 optimizations, each to be invoked with 4 different, fixed parameters).

The next class of optimizations are assembly level optimizations operating on a virtual low-level IR (see class *vLIR* in Fig. 7.3). The optimizations are listed in columns labeled with *Virtual* in Table 3.2 on p. 32. Including the parametric loop invariant code motion, which provides a conservative and an aggressive strategy, the total number of considered optimizations amounts to 7.

Following the workflow in Fig. 7.3, the register allocation is applied next. This step can be considered as an optimization, but unlike other optimizations, the application of the allocator is not optional but mandatory in order to generate valid code. The adaptive WCC supports two different register allocators which can be freely selected. Thus, the register allocator is part of the optimization sequence that can be constructed by the search algorithm. WCC implements a standard graph coloring based register allocation [Bri92] and a parametric optimal allocation [GW96] (using two different allocation strategies) leading to three different choices in total for this optimization class. It should be mentioned that this is the first work that considers different register allocators during an iterative compilation.

Finally, a local instruction scheduling can be applied as assembly level optimization on the physical low-level IR, belonging to the class *pLIR*. The second *pLIR* optimization, the generation of 16 bit instructions is not optional but always applied by the WCC, hence it could be not considered for the iterative compilation.

It should be noted that all optimizations listed in Table 3.2 are applied in WCC's standard optimization level *O1*. An exception is the register allocation, which must always be applied, and local instruction scheduling that is activated in *O2*. Also note that the order of the optimizations in the Tables 3.1 and 3.2 does not correspond to the order in which the optimizations are performed by WCC in the particular optimization levels.

In addition to the discussed optimization classes, each class (except the register allocation) contains a *dummy* optimization that, if passed to the compiler, has no effect on the code. This dummy is used to address the shortcomings of genetic algorithms that might show poor performance resulting from fixed-sequence length limitations [GSC07]. By including an arbitrary number of dummy optimizations in each sequence, the sequence length can be considered variable w.r.t. the applied optimizations.

7.4.3 Encoding of Optimization Sequences

According to Fig. 7.2, the search algorithm maintains a population of optimization sequences. After compiling the code, the objectives are determined and depending on their values, some sequences are selected for the next generation. Since an exhaustive search of all optimizations is not feasible, evolutionary optimizations have shown to be a suitable solution to this problem [KHW+05]. In this work, a particular class of evolutionary algorithms, namely genetic algorithms, is utilized. Genetic

Fig. 7.5 Encoding of optimization sequences

Fig. 7.6 One-point crossover reproduction

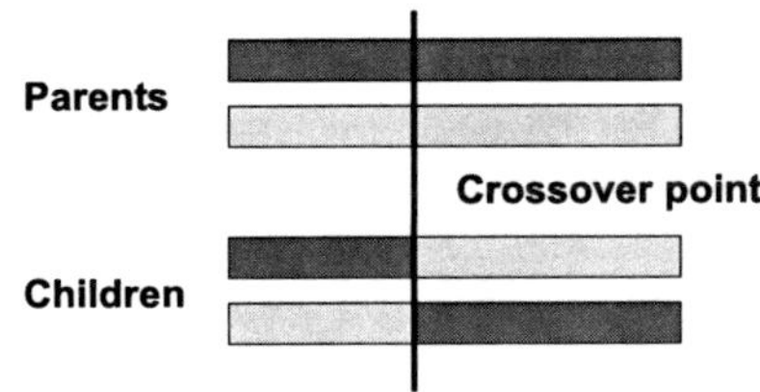

algorithms operate on strings, called chromosomes, to represent their candidate solutions. The strings may for example consist of a series of bits, characters, or integer values.

The phase ordering naturally maps to the notion of a chromosome and can be easily encoded as a string where each character denotes a specific optimization. This problem representation can be easily handled by the genetic algorithm operators *mutation* and *crossover*. Inspired by biological evolution, these operators randomly modify chromosomes to create the next generation. For more details on genetic algorithms, the interested reader is referred to standard literature [Hol92].

Figure 7.5 shows a possible encoding for the IR classes of optimizations when all optimizations are chosen in a sorted order. The numbers in parentheses denote the number of optimizations (either *real* or dummy optimization) in the respective fraction of the chromosome. For the class *register allocation (RA)* only one optimization is encoded. In the other classes, each optimization as well as the dummy optimization is encoded by a separate character.

Based on this data, WCC's compiler optimization level search space consists of $31^{31} * 8^8 * 3 * 2$ (in the order of $\approx 10^{53}$) possible permutations. This huge number emphasizes that an exhaustive search is beyond any feasible computation.

Using this string encoding, the algorithm for the *one-character mutation* operator works as follows:

1. Randomly choose the optimization IR class $Class \in \{HIR, vLIR, RA, pLIR\}$
2. In *Class*, choose a character c at a random position
3. Replace c by a character $c' \in Class$, with $c \neq c'$

Note that mutation might result in sequences where the same character (optimization) occurs multiple times in the string. Such optimization sequences are intended since equal optimizations applied at different positions in the optimization chain might have a different impact on the code, thus such sequences represent unequal individuals.

The second operator *one-point crossover* is performed in a standard, well known manner by swapping two strings at a randomly chosen position. The reproduction of two parent chromosomes using a single crossover point is depicted in Fig. 7.6.

The randomness in the evolutionary operators ensures that the genetic algorithm does not get stuck in locally optimal solutions and is likely to reach the global optimum if it is run for a sufficient number of generations.

In summary, the goal of the genetic algorithm to create optimization sequences is twofold. The algorithm

- specifies which optimizations are included in each sequence and whether some positions in the optimization sequences are filled with dummy optimizations having no effect on the code.
- defines for each sequence the order of performed optimizations in each class. In contrast to WCC's standard optimization levels, the order inside each IR class is arbitrary.

7.4.4 Objective Functions

During iterative compilation, the search algorithm requires information about the quantified objectives when a particular optimization sequence is applied. This information serves as *fitness* function to the evolutionary algorithm. As this work targets the improvement of the worst-case performance of real-time systems, the WCET has to be estimated for each generated code using the respective optimization sequence. This objective is provided by the static WCET analyzer which is tightly integrated into WCC. Further objectives involved in the optimization sequence exploration are the program's ACET and the resulting code size. The ACET is determined by an instruction set simulator, while the code size can be easily extracted from the assembly code.

To accelerate the evaluation of different objectives, maps are utilized which hold the evaluated objective values for each considered optimization sequence. Whenever an objective of a sequence has to be determined, that was already evaluated in the past, a costly re-evaluation is omitted and the objective value is efficiently obtained from the map.

7.5 Multi-objective Exploration of Compiler Optimizations

The previous section discussed the adaptive WCC framework that can be involved in compiler optimization sequence exploration. Due to the huge search space, evolutionary search algorithms are used. To satisfy the requirements for the design of modern embedded real-time systems, multiple objectives have to be considered in concert. In this section, the multi-objective exploration of compiler optimization sequences is discussed.

Section 7.5.1 introduces the general idea behind the multi-objective optimization and defines basic terms. The main characteristics of different popular *evolutionary multi-objective optimization (EMO) algorithms* applied in this study are briefly presented in Sect. 7.5.2. Since the comparison of the quality of EMO algorithms is not trivial due to their stochastic nature, a performance assessment based on statistical methods is performed. The results help to find the algorithm that is best suited for a specific problem. Principles of the performance assessment are provided in Sect. 7.5.3.

7.5.1 Multi-objective Optimization

In many real-life problems, various objectives exhibit conflicts. In case of code generation for embedded real-time systems, a trade-off between the WCET, ACET, and code size has to be taken into account. As a consequence, optimizing the application w.r.t. a single objective might yield unacceptable results for other objectives, thus an ideal multi-objective solution, which simultaneously optimizes each objective, does not exist. To cope with this problem, a set of solutions is determined. These solutions have the characteristics that on the one hand, each solution satisfies the objectives at a tolerable level and on the other hand, none of the solutions is dominated by another solution. Solutions meeting these characteristics are called *Pareto optimal* solutions.

7.5.1.1 Pareto Front Approximation

Without loss of generality, it should be assumed that all objectives are to be minimized. A translation into a maximization problem can be easily achieved by multiplying the objectives by -1. Pareto optimality, dominance, and Pareto sets are formally defined as follows [LTZ+02]:

Definition 7.1 (Pareto optimality, dominance, Pareto set) Let X denote the decision space (or search space), Z represents the objective space, $f : X \rightarrow Z$ is a function that assigns each decision vector $x \in X$ a corresponding objective vector $z = f(x) \in Z$, and m denotes the number of objectives under consideration. A decision vector $x^* \in X$ is *Pareto optimal* iff there is no other $x \in X$ that dominates x^*. x *dominates* x^*, denoted as $x \succ x^*$, iff $f_i(x) \leq f_i(x^*)$, $\forall i = 1, \ldots, m$ and $f_i(x) < f_i(x^*)$ for at least one index i. The set of all Pareto optimal decision vectors X^* is called *Pareto set*.

In other words, the decision vectors of the Pareto set can not be improved w.r.t. any other objective function without worsening at least one of the other objectives. Based on Definition 7.1, the *Pareto front* is defined as follows:

Definition 7.2 (Pareto front) Let X^* be a Pareto set. $F^* = f(X^*)$ is the set of all Pareto optimal objective vectors and is denoted as the *Pareto front*.

Fig. 7.7 Pareto fronts

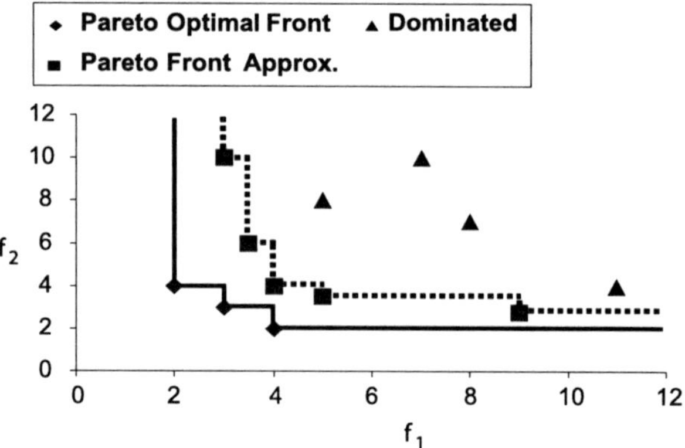

In practice, the generation of a set of decision vectors representing the entire Pareto front is often infeasible due to several reasons. For example, the number of Pareto optimal decision vectors may be too large and even the determination of a single Pareto optimum may be $\mathcal{NP}$-hard [KTZ05]. Therefore, the goal is to find a *Pareto front approximation* that minimally deviates from the Pareto optimal front. The relationship between a Pareto optimal front, its approximation, and dominated solutions for a minimization problem involving two objective functions f_1 and f_2 is depicted in Fig. 7.7.

In the compiler domain, the Pareto front approximation can be used for two different purposes. On the one hand, it helps compiler writers to find suitable optimization levels. By constructing an approximation set for a large number of benchmarks, particular points from this set that satisfy given trade-offs between the considered objectives can be chosen. The optimization sequences that represent these points may be implemented as optimization levels into the compiler and can be used in the future for new applications. This is also the scenario that is pursued in this work. On the other hand, Pareto front approximations can be exploited by compiler users to tune the optimization sequence towards a single application. In contrast to the construction of optimization levels, the approximation set is computed for a single application and the most suitable solutions are selected.

7.5.2 Evolutionary Multi-objective Optimization Algorithms

In the past, it was shown that randomized evolutionary multi-objective optimization (EMO) algorithms are best suited for the approximation of Pareto fronts [KCS06]. The algorithms basically differ in the fitness assignment, their strategy to maintain elitist solutions which will survive in the next generation, and their promotion of diversity, i.e., if a uniform distribution of solutions over the Pareto front can be attained.

The goal of this study is to find the evolutionary multi-objective algorithm that performs best for WCC's phase ordering problem. Other works [HE08] studying the impact of multi-objective optimizations in the context of iterative compilation exclusively explored a *single* EMO algorithm. Thus, it is not clear if the selected algorithm is suitable or if another optimizer would perform better in this problem domain. To cover a broad spectrum of principles used by evolutionary algorithms, a large study was conducted where three popular and credible algorithms, which have been exploited for different application domains in the past, are evaluated. These state-of-the-art optimization algorithms were chosen since each of them exhibits a different functionality. In the following, each algorithm will be briefly introduced and its specific features will be pointed out.

1. **Indicator Based Evolutionary Algorithm:**

   In contrast to other algorithms, the Indicator Based Evolutionary Algorithm (*IBEA*) [ZK04] determines fitness values by comparing individuals based on binary performance measures (called *indicators*) such as the additive ϵ-indicator. This technique has two advantages. First, the algorithm can be adapted to the user's preferences. Second, a preservation of the diversity of solutions is not required.

2. **Non-dominated Sorting Genetic Algorithm 2:**

   The fitness assignment of the computationally fast Non-dominated Sorting Genetic Algorithm (*NSGA-II*) [DAPM00] involves a non-dominated sorting of individuals. Besides the low computational requirements, NSGA-II is a parameterless approach. The only parameter defines the number of best solutions that a selection operator determines from a mating pool combining the parent and child population.

3. **Strength Pareto Evolutionary Algorithm 2:**

   The elitist Strength Pareto Evolutionary Algorithm 2 (*SPEA-2*) [ZLT01] has three main characteristics. First, the fitness assignment for each individual i in the archive (set of Pareto solutions among all so far considered generations, used for creation of new generations) is based on the number of solutions that i dominates. Second, the density estimation utilizes a k-th nearest neighbor method. Finally, an enhanced archive truncation method ensures that extreme points are preserved in the solution space.

7.5.3 *Statistical Performance Assessment*

The typical dilemma with multi-objective optimizations is indicated in Fig. 7.8. As shown on the left-hand side, numerous problem-specific modules exist. These modules serve as a representation of the problem as well as for the evaluation and variation of the solutions. An example is a compiler or a design space exploration for network processor architectures [TCG+02]. Due to their purpose, these modules are often called *variators*. To approximate Pareto optimal solutions, each of these modules can be arbitrarily combined with any evolutionary multi-objective optimization algorithm.

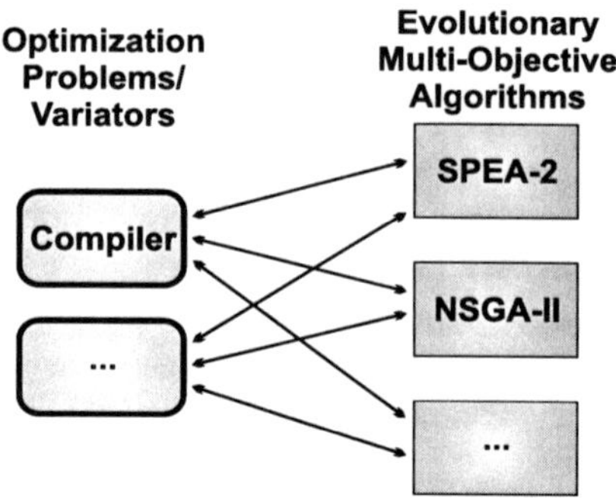

Fig. 7.8 Combining optimization problems and algorithms

While an algorithm expert is interested in the performance of his novel optimizer on real-life problems, application engineers are looking for an EMO algorithm that generates best results for their specific problem. Typically, each user group represents experts of their own domain, lacking an in-depth knowledge of the other field. Thus, it is good practice to separate the optimization problem from the algorithm and allow arbitrary combinations of both parts for an independent performance evaluation.

For the exploration of compiler optimization sequences, a manual combination and evaluation of WCC and the discussed EMO algorithms is time-consuming and error-prone. Moreover, a reliable comparison of the quality of the stochastic multi-objective optimizers is not trivial. An example are crossing Pareto fronts where a visual comparison is not intuitive anymore. To this end, an automatic and reliable performance assessment is required.

Since many EMO algorithms (including those that are considered in this study) are based on a randomized search, a simple comparison of the approximated Pareto optimal solutions generated for a specific seed is not sufficient to judge about the optimizers' performance. In order to deal with the stochastic nature of the algorithms, each algorithm has to be run multiple times for each problem with different random seeds to generate a sample of different Pareto approximation sets. These sets can be statically analyzed, i.e., a statistical hypothesis testing can be conducted to indicate if the results are *significantly different* [KTZ05]. A result is considered significantly different if it is unlikely that it occurred by chance.

To deal with the fact that complete Pareto fronts can not be determined but have to be approximated, it is common practice to assume that the given data is consistent with a simpler explanation, the so-called *null hypothesis* H_0, and then to test how likely this assumption is. A possible null hypothesis could be of the form "sets A and B are drawn from the same distribution". Using statistical hypothesis testing, the assumed null hypothesis can be rejected, indicating that there is a likely difference between A and B. For this purpose, statistical testing computes the *p-value*. This value represents the probability of obtaining a difference between two sets which is at least as large as the one that was actually observed, assuming the null hypothesis is true. The lower the p-value, the less likely the result.

In general, one rejects the null hypothesis if the p-value is smaller than or equal to a specific threshold, known in literature as the *significance level* α. A widely

---

Algorithm 7.1 Statistical hypothesis testing

Input: Significance level α
Output: Assumption about hypotheses
 1: Formulate the null hypothesis H_0 and the alternative hypothesis H_A
 2: Determine test statistics that allow an assessment of the truth of H_0
 3: Use appropriate statistical testing for assumed hypotheses to compute the p-value
 4: $result = \begin{cases} reject\ H_0\ in\ favor\ of\ H_A, & \text{if p-value} \le \alpha \\ accept\ H_0, & \text{otherwise} \end{cases}$
 5: **return** $result$

---

used value for α is 5%. Using this level, results that are only 5% likely or less are considered extraordinary. Since the null hypothesis indicates whether it is likely that two approximation sets are equal, an opposite hypothesis, the *alternative hypothesis* H_A, can be implied in addition. Using this hypothesis, the preference of samples can be tested. For example, the alternative hypothesis may state that "sample A comes from a better distribution than B".

The alternative hypothesis is taken to be true (accepted) if and only if the null hypothesis is rejected. Which alternative hypothesis is implied, depends on the statistical test. For example, the *Mann-Whitney* rank sum test [Con71], which is conducted in this book, considers samples pairwise: $test(A, B)$. The underlying alternative hypothesis is that A is better than B.

In general, hypothesis testing consists of four steps as shown in Algorithm 7.1. *Test statistics* used in the second step designate an appropriate numerical summary of the evaluated data sets, i.e., the sets are reduced to a single or a small number of representative values. The following example demonstrates this approach.

Example 7.1 It should be tested whether the Pareto front approximation set generated by algorithm A is better than that of algorithm B. This statement is reflected by the alternative hypothesis. The assumed null hypothesis states that none of the sets is better than the other. Using statistical hypothesis testing, the following p-values for the test statistics T_A and T_B for sets A and B, respectively, are computed:

$$test(T_A, T_B) = 0.002 \land test(T_B, T_A) = 0.998$$

With respect to the significance level of 5% ($\alpha = 0.05$), the null hypothesis is rejected for the first test (p-value $= 0.002$) in favor of the alternative hypothesis. Hence, the sets significantly differ and the results suggest that A may be better than B.

As can be seen in Algorithm 7.1, the assessment of statistical significance of the considered EMO algorithms relies on test statistics. In the following, two different approaches for the computation of test statistics are presented.

7.5.3.1 Dominance Ranking

The first approach is based on *dominance ranking* [KTZ05]. Its main idea is to rank the points of the approximation sets based on the dominance relation. To this end, for each considered EMO algorithm $i \in \{1, \ldots, q\}$, a number of runs $r_i \geq 1$ is performed, generating a collection C of approximation sets $A_1^1, A_2^1, \ldots, A_{r_1}^1, \ldots, A_1^q, \ldots, A_{r_q}^q$. From this collection C, each set z is assigned a rank based on the dominance relations by counting the number of sets which are *better* than z. The dominance relation *better* ($\lhd$) in the context of approximation sets is defined as follows:

Definition 7.3 (Better relation $\lhd$) Set A is better than B, denoted as $A \lhd B$, iff every $z_2 \in B$ is weakly dominated by at least one $z_1 \in A$ and A is different from B. z_1 weakly dominates z_2 iff z_1 is not worse than z_2 in all objectives.

The rank of each approximation set contained in the collection C is computed as follows:

$$rank(C_i) = 1 + |\{C_j \in C : C_j \lhd C_i\}|$$

The lower the rank, the better the corresponding approximation set with respect to the entire collection. Using these results, each approximation set can be reduced to a single number, representing the respective test statistics. These statistics are utilized in step 2 of Algorithm 7.1 in order to determine whether these values significantly differ, i.e., whether the ranks for one algorithm are significantly smaller than the ranks for another EMO algorithm.

Using dominance ranking, the question whether one algorithm performs better than another can be answered. However, its disadvantage is that no quantitative statements about the difference in quality of the EMO algorithms can be made. Therefore, it is recommended to perform complementary tests for the performance assessment that allow a quantification of the results in order to further characterize the differences in the approximation set distributions [KTZ05]. For this purpose, the second approach of WCC's statistical testing exploits *quality indicators*.

7.5.3.2 Hypervolume Indicators

A quality indicator I is a mapping from the set of all approximation sets Ω to the set of real numbers:

$$I : \Omega \mapsto \mathbb{R}$$

In contrast to dominance ranking, the indicators provide a quantitative measure. That is, given two approximation sets A and B, a comparison of the indicator values $I(A)$ and $I(B)$ allows to draw the conclusion to which extent one algorithm outperforms another.

A popular quality indicator, which is also used in this study, is the *hypervolume indicator* I_H [ZT99]. This indicator computes the hypervolume of the portion of

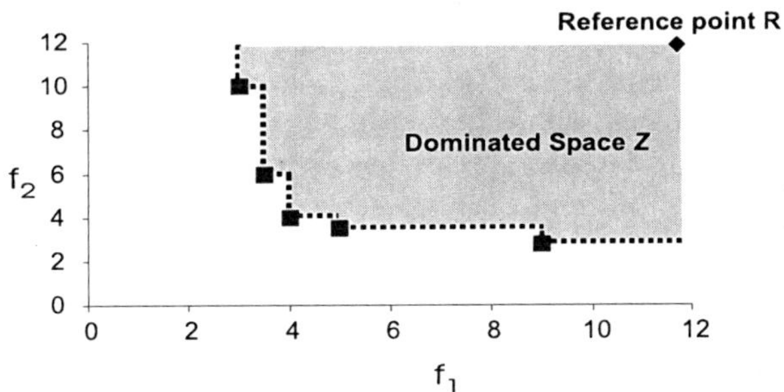

Fig. 7.9 Example for hypervolume indicator

the objective space Z that is weakly dominated by an approximation set A, i.e., any objective vector in Z is weakly dominated by at least one objective vector in A. To measure the hypervolume, the objective space is bounded by a *reference point* R that has to be weakly dominated by all objective vectors in A. The hypervolume indicator is illustrated in Fig. 7.9.

Larger values of the hypervolume indicator correspond to higher quality. Thus, for two sets A and B, the result $I_H(A) > I_H(B)$ implies that A is better than B with respect to the hypervolume indicator. The difference of the hypervolume indicators quantify the difference in the quality of the sets.

However, this information comes at the cost of generality. In contrast to dominance ranking, the results of quality indicators represent specific assumptions about the preferences of the decision maker, called *preference information*. Hence, these indicators should be used as a complementary technique to dominance ranking which yields statements independent of any preference information.

Computation of the hypervolume indicator has been shown to be exponential in the number of objectives and polynomial in the number of objective vectors in the approximation sets [KTZ05]. For the considered problem of the optimization sequence exploration with typically small approximation sets, the optimization run time of this indicator is negligible.

To address the stochastic nature of the EMO algorithms, the capability of quality indicators to reduce the dimension of an approximation set to a single value is exploited. Similar to dominance raking, the algorithms are run with different random seeds and the generated approximation sets are collected into a set C^I. Based on this collection, an appropriate reference point is selected and the hypervolume indicators are computed. Finally, statistical testing is conducted on these test statistics.

7.5.3.3 Statistical Hypothesis Testing

WCC's performance assessment of the stochastic multi-objective optimizers is based on recommendations proposed in [KTZ05]. Since the results of a single assessment method are often not sufficient, the two previously mentioned approaches are applied in concert and their results are related to each other in order to obtain reliable performance statements. The following steps are carried out:

1. **Preprocessing**: The computed approximation sets of the three considered EMO algorithms are collected for runs with different seeds. Based on this collection C, the lower and upper bounds of the objective vectors are computed. The bounds are used to normalize all objective vectors, such that all values lie in the interval $[1, 2]$. Moreover, based on all Pareto solutions in C, a non-dominated front of objective vectors is determined, serving as *reference set* for subsequent steps. The reference set can be seen as an overall Pareto front approximation considered for all optimizer results.
2. **Dominance ranking**: For each normalized optimizer approximation set, the dominance ranking procedure is applied for each pair of the considered EMO algorithms. The determined test statistics are assessed using the Mann-Whitney test (cf. step 2 in Algorithm 7.1) to compute the p-value.
3. **Application of the hypervolume indicators**: For each normalized approximation set and the reference set, the quality indicator hypervolume is computed for each pair of the EMO algorithms. The results are visualized to allow a convenient evaluation. Furthermore, statistical tests (cf. Algorithm 7.1) are carried out based on the Mann-Whitney test.

7.6 Experimental Results for Optimization Exploration

To indicate the efficacy of the found multi-objective optimization sequences, evaluation on a large number of different real-life benchmarks was performed. Similar to the evaluation of the machine learning based heuristics in Chap. 6, the experiments were conducted as *cross validation*. One set of benchmarks, the *training set*, was used during the multi-objective search. The determined sequences are subsequently evaluated on unseen benchmarks, representing the *test set*. This approach enables an estimation of the generalization ability, i.e., the results suggest which improvements of the objective functions can be expected for unseen programs. Moreover, the cross validation conforms with the ideology of standard compiler optimization levels which are developed to perform well on a large number of future programs.

As in the previous chapters, the benchmarks stem from DSPstone [ZVS+94], MediaBench [LPMS97], MiBench [GRE+01], MRTC [MWRG10], NetBench [MMSH01], and UTDSP [UTD10]. The large variety of benchmark suites emphasizes WCC's focus on generality. By covering a large number of different embedded systems applications during the search, future embedded software should benefit in a similar fashion from WCC's novel optimization sequences. For the present study, the training and test set each consisted of 35 benchmarks, exhibiting a similar distribution, i.e., from each benchmark suite approximately the same number of benchmarks was selected for each set.

The fully automated workflow of the employed framework for the multi-objective exploration of compiler optimizations is depicted in Fig. 7.10. The process starts with the variator that generates random compiler optimization sequences, representing the initial population. The optimizations are encoded as strings. The

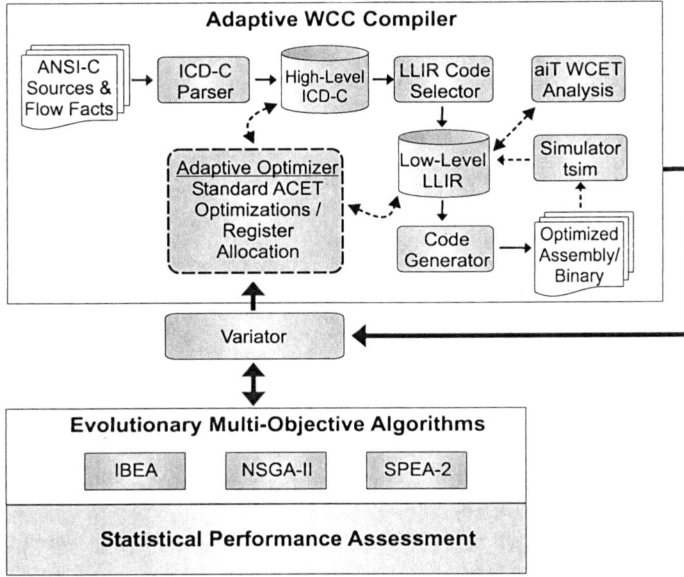

Fig. 7.10 Adaptive WCC for multi-objective compiler optimization exploration

sequences are passed to the adaptive WCC framework which is extended by an adaptive optimizer that can perform code transformations in an arbitrary order (cf. Sect. 7.4.2). WCC uses these sequences to generate code which is processed by the WCET analyzer aiT and the TriCore simulator *tsim*. The estimated WCET, ACET, and code size (obtained from LLIR) are returned to the variator. This way, the variator manages for each optimization sequence the corresponding objective values. For distinction, each sequence is indexed using a unique ID.

In the next step, for each evaluated optimization sequence of a particular generation, the variator passes the corresponding indices and objective vectors to the evolutionary multi-objective algorithms. Exclusively based on the objective vectors, the EMO algorithms compute the approximated Pareto front and return the respective indices of the Pareto solutions back to the variator. The returned solutions are finally used by the mutation and crossover operators to generate individuals for the next generation of optimization sequences. Moreover, the computed Pareto front approximations serve as input to the statistical performance assessment.

Both, the collection of the EMO algorithms and the performance assessment are part of the *Platform and Programming Language Independent Interface for Search Algorithms (PISA)* framework [BLTZ03]. An existing PISA variator was re-implemented in order to be applicable with the WCC framework.

For the conducted experiments, the following parameters were used:

- the algorithms IBEA, NSGA-II, and SPEA-2 were run 5 times with different random seeds
- for each run, each population comprises 50 individuals (optimization sequences)
- the archive [BLTZ03] holds 25 individuals

Table 7.1 Dominance ranking results for ⟨WCET, ACET⟩ and ⟨WCET, Code Size⟩ using Mann-Whitney rank sum test

	⟨WCET, ACET⟩			⟨WCET, Code Size⟩		
	IBEA	**NSGA-II**	**SPEA2**	**IBEA**	**NSGA-II**	**SPEA2**
IBEA	–	0.760	0.949	–	0.5	0.5
NSGA-II	0.240	–	0.011	0.5	–	0.016
SPEA2	0.051	0.899	–	0.5	0.984	–

- a one-character mutation (cf. Sect. 7.4.3) with a probability of 10%
- a one-point cross-over probability of 90%
- optimization was run for 50 generations
- statistical performance assessment with a significance level $\alpha = 5\%$

These are common settings for the genetic algorithms used for the exploration of the compiler optimization space [ZKW+04, HE08].

In the following, the Pareto front approximations found by the three considered EMO algorithms are evaluated. The multi-objective optimizations are carried out for the following pairs of objectives: ⟨WCET, ACET⟩ and ⟨WCET, code size⟩. The consideration of 2-dimensional Pareto fronts is motivated by two issues:

- Since the impact of standard optimizations is unknown so far for the trade-off between the WCET and other objectives, this book is the first case study to investigate this issue. The results help to understand the basic interferences between different objectives. Starting with the investigation of more than two objectives may hide some objective interferences, leading to a lack of the fundamental understanding.
- Compiler writers typically consider a trade-off between two objective functions, thus the results of this work are more valuable for them than the presentation of complex, often not intuitive, objective interferences.

7.6.1 Statistical Performance Assessment

Table 7.1 presents results for dominance ranking computed by the *Mann-Whitney* rank sum test for the possible combinations of the considered algorithms IBEA, NSGA-II, and SPEA2. Columns 2–4 indicate p-values for the objective functions ⟨WCET, ACET⟩, while columns 5–7 present p-values for the objectives ⟨WCET, code size⟩.

The statistical tests are performed pairwise w.r.t. the alternative hypothesis [KTZ05] that the dominance ranks for the algorithms in the first column are significantly better than those for the algorithm in the following columns.

For the objectives ⟨WCET, ACET⟩, the p-value of 0.011 in the fourth row and fourth column denotes that the difference between NSGA-II and SPEA2 is signifi-

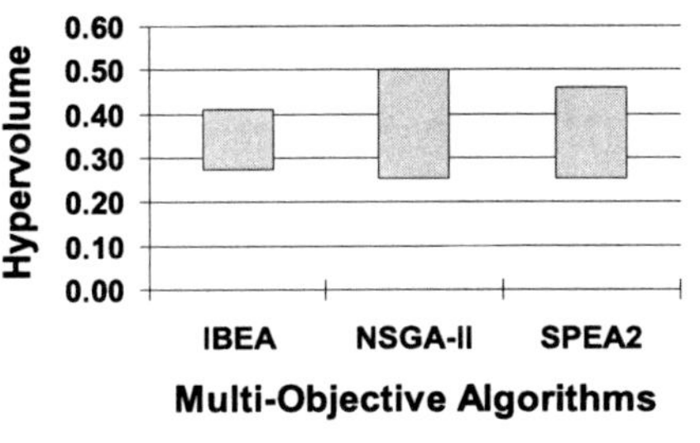

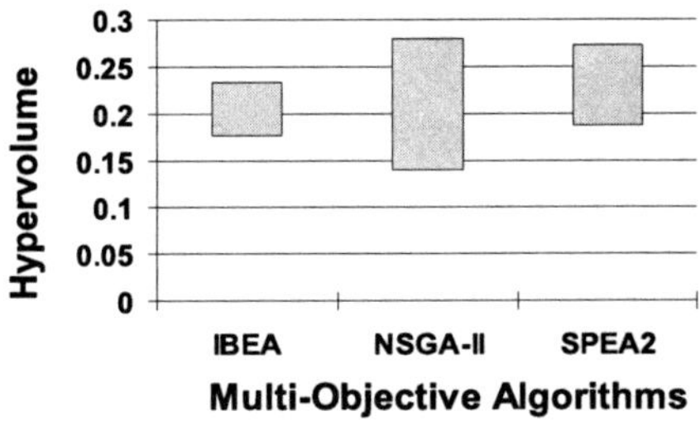

(a) Objective Pair ⟨WCET, ACET⟩ (b) Objective Pair ⟨WCET, Code Size⟩

Fig. 7.11 Distribution of hypervolume indicator for different objective pairs

Table 7.2 Hypervolume indicator results for ⟨WCET, ACET⟩ and ⟨WCET, Code Size⟩ using Mann-Whitney rank sum test

	⟨WCET, ACET⟩			⟨WCET, Code Size⟩		
	IBEA	**NSGA-II**	**SPEA2**	**IBEA**	**NSGA-II**	**SPEA2**
IBEA	–	0.899	0.889	–	0.961	0.803
NSGA-II	0.011	–	0.046	0.039	–	0.041
SPEA2	0.111	0.954	–	0.197	0.959	–

cant, implying that NSGA-II outperforms SPEA2. For other optimizer pairs, no significant differences were observed. The results for the objectives ⟨WCET, code size⟩ lead to the same conclusion. NSGA-II outperforms SPEA2 since the dominance ranking results significantly differ (p-value $= 0.016$) w.r.t. $\alpha = 0.05$. Hence, NSGA-II seems to be the most promising EMO for the given problems. There are also differences between other combinations of the algorithms but they are not significant.

The distribution of the results of the hypervolume indicator for different objective pairs and different algorithms is depicted as boxes in Fig. 7.11, showing 50% of the results around the median. In Fig. 7.11(a), the results for the objective pair ⟨WCET, ACET⟩ indicate that NSGA-II covers a larger hypervolume than IBEA and SPEA2 (box capacity is larger). Thus, it can be assumed that for the conducted experiments, NSGA-II computes best optimization sequences exhibiting the highest performance. Comparable results are achieved for the objective pair ⟨WCET, code size⟩, as can be seen in Fig. 7.11(b). The results provide a further hint that the NSGA-II algorithm outperforms IBEA and SPEA2 for the search of the compiler optimization space. However, to validate this assumption, statistical significance testing is required.

To test statistical significance of the quality indicator, Table 7.2 shows results for the indicators' hypervolume utilizing the statistical Mann-Whitney tests computed for the objective pairs ⟨WCET, ACET⟩ and ⟨WCET, code size⟩. Similar to the dominance ranking test, the p-values in the table are computed w.r.t. the alternative hypothesis that the indicator results for the algorithms in the first column are significantly better than those for the algorithm in the following columns. It can be

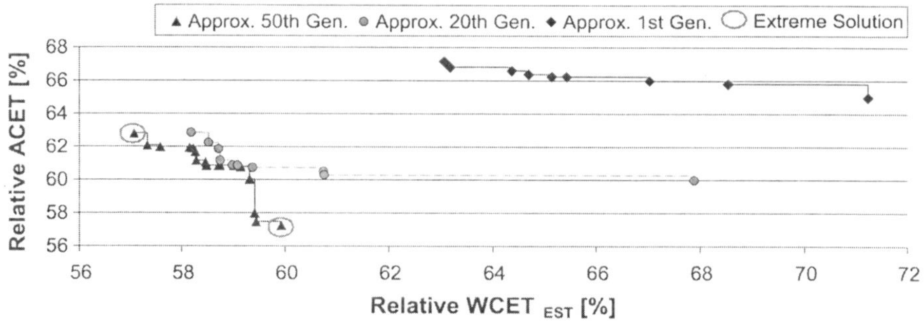

Fig. 7.12 NSGA-II Pareto front approximation for ⟨WCET, ACET⟩

observed that there are statistically significant differences for the significance level of 5% using the hypervolume indicator.

For the objectives ⟨WCET, ACET⟩, NSGA-II outperforms IBEA (p-value = 0.011) and SPEA2 (p-value = 0.046). For other pairwise tests, no statistical significances was discovered. Compared to the dominance ranking results, which also signified a preference of the NSGA-II algorithm, NSGA-II seems to be the best choice for the search of Pareto optimal sequences for the objectives WCET and ACET. These results conform with results reported in [KBT+04] where the authors observed that NSGA-II typically outperforms SPEA2 for problems with two objectives.

Comparable results for the statistical significance testing of the hypervolume indicator are achieved for ⟨WCET, code size⟩. As can be seen in the last three columns of Table 7.2, NSGA-II outperforms IBEA and SPEA-II since the respective p-values of 0.039 and 0.041 are smaller than the significance level, thus allowing to reject the null hypothesis in favor of the alternative hypothesis. Hence, NSGA-II also promises to be the best algorithm for the optimization sequence exploration w.r.t. the WCET and code size.

7.6.2 Analysis of Pareto Front Approximations

Figure 7.12 visualizes the Pareto front approximations for the training set generated by the algorithm NSGA-II which achieved best performance assessment results for the objective functions WCET and ACET. The front approximations are depicted for the 1st, 20th, and 50th generation and contain the accumulated Pareto solutions of the 5 runs of the algorithm with different random seeds. The final front for the 50th generation consists of 19 Pareto solutions. The horizontal axis indicates the relative WCET estimation w.r.t. the non-optimized code, i.e., 100% represents the WCET with all disabled WCC optimizations. In a similar fashion, the vertical axis represents the relative ACET w.r.t. to the non-optimized code. Based on this figure, the following can be concluded:

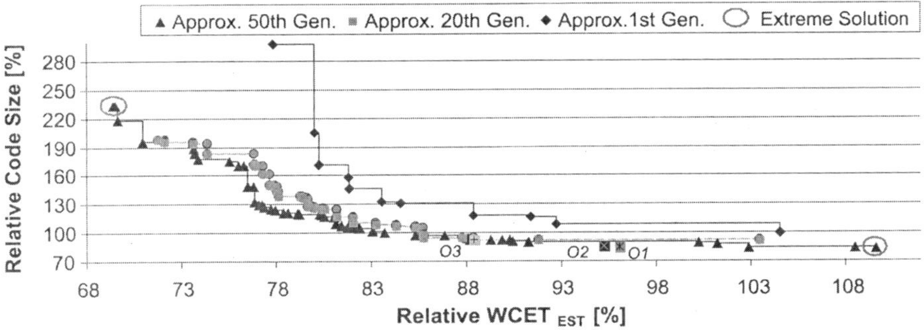

Fig. 7.13 NSGA-II Pareto front approximation for ⟨WCET, code size⟩

- It is worthwhile to invest time in the evolutionary search. While the first generation achieves average WCET and ACET reductions of 36.9% and 35.0%, respectively, for all benchmarks of the training set, the 50th generation reduces the WCET and ACET of up to 42.9% and 42.8%, respectively.
- The discovered sequences significantly outperform the standard optimization levels, having the (*WCET*, *ACET*)-coordinates (due to space constraints not included in the figure) (96.0, 89.1) for *O1*, (95.2, 90.4) for *O2*, and (88.4, 84.7) for *O3*. For example, the performance of *O3* is outperformed by 31.3% and 27.5% for WCET and ACET, respectively, when the *extreme* situations (circled in Fig. 7.12) are selected.
- Standard compiler optimizations have a similar impact on the WCET and ACET. This observation provides an important answer to the question which concerns all designers of real-time systems: *which impact can be expected from standard ACET optimizations on the system's worst-case behavior?* This case study shows that similar effects on the average-case and worst-case behavior are likely.
- Even though the effect of standard optimizations on the WCET and ACET is comparable, the results emphasize the importance of the development of WCET-aware optimizations. If a maximal WCET reduction is desired, novel optimizations are required that focus on an aggressive WCET reduction at the cost of a degraded ACET.

The Pareto front approximation computed by NSGA-II for the objectives WCET and code size is depicted in Fig. 7.13. The relative WCET estimation w.r.t. the non-optimized code (corresponds to 100%) is represented by the horizontal axis, while the relative code size w.r.t. to the non-optimized code is shown on the vertical axis. Again, Pareto front approximations of the 1st, 20th, and 50th generation are visualized and are constructed of Pareto solutions found in the 5 runs of the algorithm. The 50th-generation front comprises 53 points. Compared to the Pareto front approximations for ⟨WCET, ACET⟩, the interpretation equals in two points:

- The evolutionary search pays off for both objective functions. For the first generation, a WCET reduction of 21.2% at the cost of the code size increase of 197.4% can be achieved (left-most solutions of the corresponding front). If code size is the crucial objective, a code size reduction of 0.4% with a simultaneous WCET increase of 4.5% can be observed. For the 50th generation, the following extreme solutions were observed (marked by circles): a WCET reduction of 30.6% with a simultaneous code size increase of 133.4%, or a WCET degradation of 9.6% with a simultaneous code size reduction of 16.9%. Hence, the results for later generations yield substantially better results.

- The Pareto solutions outperform WCC's standard optimization levels which are depicted in Fig. 7.13. The standard optimization levels perform well for the code size reduction. Using *O2*, which does not include code expanding optimizations, a code size reduction of 14.9% can be achieved on average, while NSGA-II reduces the code size by up to 16.9%. Moreover, WCC's maximal WCET reduction of 13.6% found by *O3* can be outperformed by the found Pareto solutions by 17.0%, amounting to a WCET reduction of 30.6% as found by NSGA-II.

However, there are also two major differences compared to the results of the objective pair ⟨WCET, ACET⟩. The WCET and the code size are typical conflicting goals. If a high improvement of one objective function is desired, a significant degradation of the other objective must be accepted. This is an important conclusion for memory-restricted real-time systems. To achieve a high WCET reduction, the system must be possibly equipped with additional memory to cope with the resulting code expansion. Also, compiler writers developing WCET-aware optimizations must be aware of these conflicting objectives and should always consider the impact of their optimizations on the code size. The second difference is that standard compiler optimizations available in WCC are not capable of accomplishing a notable code size decrease. Therefore, tailored optimizations are required if code size reduction is a primary goal.

7.6.3 Analysis of the Optimization Sequences

A closer look at the Pareto optimal optimization sequences for the objective pair ⟨WCET, ACET⟩ reveals the following observations:

- Most of the optimization sequences contain an aggressive loop unrolling or function inlining in the very beginning. Aggressive means that loops/callees with a maximal size of 200 expressions (maximal parameter value considered during exploration) were transformed. This observation conforms with results of Sects. 4.5 and 6.4 where it was shown that these optimizations are enabling optimizations.
- In addition to these two optimizations, the found Pareto sequences often contain the optimization procedure cloning. Since cloning, unrolling, and inlining are all contained in WCC's optimization level *O3*, it can be concluded that this standard optimization level holds promising optimizations for maximal WCET/ACET reduction.
- Other optimizations frequently found in the Pareto solutions are: instruction scheduling applied at physical LLIR, ILP-based register allocation (hence the optimization's complexity pays off), and loop-invariant code motion (the potential of this important optimization was studied in Sect. 6.5).

- Optimizations that were infrequently contained in the sequences—hence can be considered less beneficial—are: instruction scheduling applied at virtual LLIR, loop collapsing, life range splitting, and loop deindexing (cf. Sect. 7.2).
- There is no clear separation which optimizations are best suitable for a particular objective, hence in general many optimizations have a comparable effect on the estimated WCET and ACET.

The analysis of the Pareto optimal optimization sequences for the objective pair ⟨WCET, code size⟩ leads to the following observations:

- Function inlining can often be found in code size-oriented sequences. Although widely believed that the optimization always yields a code expansion, inlining can also reduce code size if functions, which are invoked once in the code, are inlined and further optimized (cf. Sect. 4.5).
- Especially for the code size-oriented solutions, many sequences begin with procedure cloning of small functions (limited to 20 expression). A possible explanation is that cloning of small function has a negligible impact on the code size but may significantly improve the estimated WCET (cf. Sect. 4.3), thus overall good Pareto solutions can be generated that way. Moreover, some of the clones can be afterwards inlined, possibly leading to a code size decrease.
- In contrast to the WCET-oriented sequences, none of the code size-oriented optimization strategies contained loop unrolling. Hence, it can be expected that unrolling is likely to yield a code size increase.
- Loop unswitching is another optimization that is rarely found in the code size-oriented sequences. As shown in Sect. 4.6.6, unswitching also increases code size, thus should be used with caution if code size is critical.
- As shown above, the objectives WCET and code size have conflicting goals. Therefore, some standard optimizations, such as peephole optimizations or dead-code elimination, have a positive effect on both objectives, while other optimizations should be applied only if improvements of one objective at the cost of the other one can be tolerated.

7.6.4 Cross Validation

To estimate the generalization ability of the discovered sequences, a cross validation was performed, i.e., optimization sequences found by NSGA-II in the 50th generation for the training set are applied to unseen benchmarks from the test set.

Among the large number of solutions constructing a Pareto front approximation, three solutions are discussed in more detail since they provide typical optimization scenarios. If the system designer is interested in a maximal reduction of a particular objective function, an optimization sequence represented by one of the extreme points from the fronts should be considered. Another alternative is to choose a Pareto solution from the middle of the front which represents a trade-off between the respective objectives.

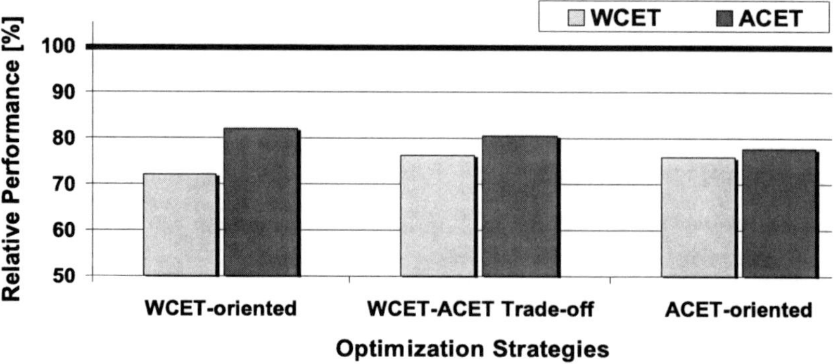

Fig. 7.14 Cross validation for ⟨WCET, ACET⟩

For the objective pair ⟨WCET, ACET⟩, the optimization sequences defined by the extreme points in Fig. 7.12 as well as the trade-off, which is represented by the solution with the coordinates (58.7, 60.9), were evaluated. These three optimization sequences were applied to each of the 35 benchmarks from the test set. For each benchmark, the results for the estimated WCET and ACET using the new sequence were compared with the WCET/ACET results when the benchmark was compiled with WCC's highest optimization level *O3*. The averaged results for all benchmarks, with 100% being the base line representing results achieved with *O3*, are shown in Fig. 7.14. Following results can be observed:

- Using the *WCET-oriented* optimization sequence, which is represented in Fig. 7.12 by the left-most Pareto solution with the *WCET, ACET* coordinate (57.1, 62.8), outperforms WCC's default optimization level *O3* by 28.0% and 18.0% for the estimated WCET and ACET reduction, respectively. It can also be seen, that the estimated WCET was improved at the cost of ACET.
- The optimization sequence (labeled with *WCET-ACET Trade-off*) was determined by NSGA-II as a compromise between the WCET estimation and ACET for the training set (see solution with coordinate (58.7, 60.9) in Fig. 7.12). For the test set, the results are slightly worse than for the training set, since a relative WCET estimation of 76.2% and a relative ACET of 80.5% were observed. However, the optimization level *O3* can still be significantly outperformed by 23.8% and 19.5% for WCET and ACET, respectively. Hence, the discovered optimization sequence is a promising candidate for the substitution of WCC's optimization level *O3*.
- The optimization sequence (coordinate (59.9, 57.2) in Fig. 7.12) aiming at the maximal ACET reduction (labeled with *ACET-oriented*) could achieve the maximal ACET reduction of 23.4% among the considered optimization strategies. This sequence has even a slightly better impact on the average WCET than on the average ACET. Hence, again *O3* can be improved.

Analogously, Fig. 7.15 reflects the performance of the sequences found for ⟨WCET, Code Size⟩ in Fig. 7.13. Using the *WCET-oriented* optimization scenario,

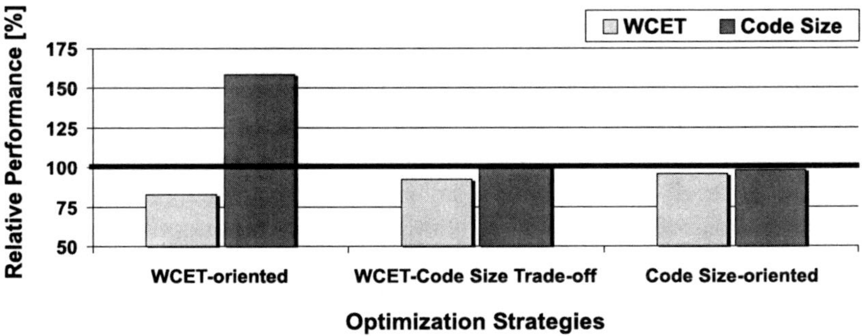

Fig. 7.15 Cross validation for (WCET, Code Size)

WCET reductions of 17.4% can be achieved on average for the test set. However, this improvement comes at the cost of a code size increase of 58.7%. The optimization sequence labeled with *WCET-Code Size Trade-off* improves the WCET by 7.8% w.r.t. *O3* but also results in a slight code size increase of 1.0%. Concerning the *Code Size-oriented* strategy, the WCET and code size can be reduced by 4.25% and 1.9%, respectively, compared to *O3*.

7.6.5 Optimization Run Time

The run time of the multi-objective exploration of optimization sequences was measured for each EMO algorithm. Five optimization runs with different seeds, each computing 50 generations, took about 6 days on a Intel Quad-Core Xeon 2.4 GHz machine with 8 GB RAM. This optimization run time might seem long. However, it should be noted that these automatic tests have to be performed once off-line, while the results (optimization sequences) can be re-used without additional overhead for a large number of devices. Therefore, the high performance requirements imposed on today's systems fully justify the observed optimization times.

The application of the new optimization sequences found by the EMO algorithm NSGA-II does not considerably increase the compilation time compared to WCC's *O3*. In some cases, like for an extensive loop unrolling or function inlining, slight increases of the compilation time (typically few seconds) could be observed.

The techniques presented in this chapter were published in [LPF+10].

7.7 Summary

Interactions between optimizations are pervasive in compilers, turning the search for promising optimization sequences into a challenging task. To relieve the user from this cumbersome search, compilers are usually equipped with standard optimization levels. However, various studies have observed that these levels may exhibit poor performance or even a performance degradation.

There are several reasons why compilers fail to deliver high performance. One reason is the application of insufficient compiler optimization heuristics that can be automatically improved by supervised machine learning, as shown in Chap. 6. Another reason is the rigid structure of today's optimizing compilers with their predefined optimization order that does not allow to adapt to the compiled programs and the underlying hardware. Moreover, the search space of compiler optimizations is too large to be manually or exhaustively explored for good optimization sequences. Besides these typical problems found in traditional compilers, embedded real-time systems impose new requirements on compilers: due to their multi-objective nature, multiple (often conflicting) objectives have to be taken into account during code generation.

This chapter tackles the three latter problems. To provide flexibility within the optimizing compiler, WCC was extended towards an adaptive compiler. This way, arbitrary optimization sequences can be generated that best adopt to the requirements of specific programs and hardware configurations. To cope with the large search space but also to address the multi-objective nature, WCC employs a multi-objective evolutionary search to find Pareto optimal optimization levels. The presented framework is fully automated and is transparent to the compiler, the applications, the hardware, and the considered objective functions. Moreover, unlike other works, the proposed approach performs a reliable assessment of the quality of the three considered stochastic multi-objective optimizers based on statistical hypothesis testing.

The search for promising optimization sequences was conducted for the objective pairs ⟨WCET, ACET⟩ and ⟨WCET, code size⟩. To analyze generality of the results, i.e., to estimate how well the identified optimization sequences will perform in the future for unseen programs, a cross validation was performed: best sequences were determined for a training set of benchmarks and afterwards evaluated on a new test set. The results clearly show that WCC's highest optimization level *O3* can be significantly outperformed when the found sequences are applied. For the test set, average WCET and ACET reductions of up to 28.0% and 23.4%, respectively, compared to *O3* were observed. Compared to the non-optimized code, this translates to an average reduction of the WCET and ACET by 37.1% and 38.1%, respectively. Considering the code size, the average code size of the benchmarks from the test set could be reduced by 1.9% compared to *O3*.

This very first comprehensive analysis of the impact of standard compiler optimizations on crucial objectives of embedded real-time systems allows to draw several conclusions. First, the results point out the relation between different objectives. The impact of standard optimizations on the worst-case and average-case performance is similar. Hence, this insight helps to clarify the real-time system designers' uncertainty about the influence of standard compiler optimizations on the program's WCET. Second, the observed code size decrease achieved with standard optimizations was marginal, emphasizing the need for tailored optimizations if a code reduction is desirable. Third, it was indicated that the WCET and code size are conflicting goals, i.e., high WCET reductions are often accompanied by a substantial code expansion. As a consequence, designers of WCET-aware optimizations should be aware of this fact and provide mechanisms to control effects on the code size.

Last but not least, valuable insights for compiler designers concerning promising optimization sequences were provided. For an effective reduction of the WCET and ACET, an aggressive inlining and unrolling, procedure cloning, ILP-based register allocation, instruction scheduling after register allocation, and loop-invariant code motion seem to be beneficial. On the other hand, optimization sequences aiming at a reduction of the WCET and code size should include a careful function inlining and procedure cloning of small functions but also exclude loop unrolling.

Chapter 8
Summary and Future Work

Contents

In this book, a compiler framework and optimization techniques for embedded real-time systems were proposed. The compiler optimizations were performed at source code and assembly level and achieved an automatic reduction of the program's worst-case execution time. Therefore, the developed techniques improve the current state of real-time system design which is typically based on a manual, time-consuming, and error-prone trial-and-error procedure. Section 8.1 summarizes the contributions of the techniques proposed in this book. Finally, a discussion on directions for future work in Sect. 8.2 concludes this work.

8.1 Research Contributions

Most embedded/cyber-physical systems have to respect stringent timing constraints. Primarily in safety-critical application domains, such as automotive or avionics, timing deadlines must be satisfied to guarantee correctness of the system. An important parameter to reason about the timeliness is the WCET that can be safely estimated by a static timing analysis.

To meet these constraints, it must be ensured that the WCET of software running on hard real-time systems does not exceed its deadlines. Within the current design process, this requirement is typically achieved by a trial-and-error procedure which implies numerous iterations of software generation and timing evaluation. Since this procedure is time-consuming and error-prone, an automatic reduction of the program's WCET is highly desirable.

P. Lokuciejewski, P. Marwedel, *Worst-Case Execution Time Aware*   229
Compilation Techniques for Real-Time Systems, Embedded Systems,
DOI 10.1007/978-90-481-9929-7_8, © Springer Science+Business Media B.V. 2011

This book addresses the lack of design tools for meeting the timing constraints of real-time systems. A novel compiler framework, the WCC, was developed which integrates timing analyses into the code generation and optimization process. Based on this reconciliation between the compiler and a WCET analyzer, numerous automatic compiler optimizations were developed that systematically improve the worst-case performance of an application at both the source code and assembly level. In addition, different static analysis techniques were presented that deliver auxiliary information for an increased optimization potential. Besides the provided automatism, these compiler optimizations are also more effective than the prevalent trial-and-error approach since their global view of the application enables the full exploitation of the optimization potential. In the following, an overview of the contributions of this work is provided in more detail.

8.1.1 Extensions to WCC Framework

These extensions to the infrastructure of the WCC framework were realized to enhance WCET-aware compilation:

Static Loop Analysis automatically determines iteration counts of loops in the application which is mandatory for timing analyses. This way, the need for tedious user-annotations is removed, turning WCC into an automated design framework.

Back-annotation translates WCET and other relevant data from the compiler backend into the frontend to enable access to this critical information in WCET-aware source code level optimizations.

Invariant Path classifies the WCEP into sub-paths where a WCEP switch can not occur. Exploiting this knowledge, WCET-aware optimizations can be accelerated since redundant updates of WCET data are eliminated.

Machine Learning techniques were reconciled into the WCC compiler framework to enable an automatic generation of compiler heuristics for WCET reduction.

Evolutionary Multi-objective Algorithms have been exploited for the generation of Pareto optimal compiler optimization sequences for the sometimes conflicting objectives WCET, ACET, and code size. As the experiments showed, the novel optimization sequences could significantly outperform WCC's highest optimization level *O3*.

8.1.2 WCET-Aware Source Code Level Optimizations

Up to now, the large potential of compiler optimizations for WCET reduction was not explored. In this book, the WCC framework with the previously mentioned extensions was utilized for the development of the following source code level optimizations:

WCET-Aware Procedure Cloning creates specialized copies of functions, making calling contexts explicit at the source code level. This way, more precise flow facts for data-dependent loops can be specified, leading to an improved WCET analysis.

WCET-Aware Superblock Optimizations operate on superblocks that consist of several basic blocks and provide more optimization opportunities than single blocks. In this book, the superblock formation was applied for the first time at source code level and was driven by WCET data. This new code structure was combined with the traditional optimizations common subexpression and dead code elimination.

WCET-Aware Loop Unrolling obtains detailed knowledge about loop iteration counts from WCC's static loop analysis. Moreover, WCC's back-annotation is involved to obtain detailed information about timing, code size and spill code. Based on this data, each loop in the program is unrolled with an individual unrolling factor that promises the highest WCET reduction.

WCET-Aware Loop Unswitching is an optimization that reduces the number of executed loop-invariant conditional branches by moving them outside the loop and copying the replicated loop bodies into the respective *then-* and *else*-parts. To avoid a too extensive code expansion but also to achieve high WCET reduction, this proposed optimization begins with those loops whose transformation promises the highest WCET reduction. Moreover, WCC's unswitching is accelerated through the use of the invariant path paradigm.

WCET-Aware Function Inlining utilizes machine learning based heuristics that predict whether inlining of the considered function call site promises a WCET reduction. This way, expensive evaluations of function inlining effects are replaced by an efficient request to a prediction model.

8.1.3 WCET-Aware Assembly Level Optimizations

As the previous chapters have shown, the exploitation of processor-specific features during code transformations may be highly beneficial. For this reason, this book proposed the following WCET-aware optimizations at the assembly level:

WCET-Aware Procedure Positioning aims at the improvement of the instruction cache behavior by re-arranging procedures in memory in such a way that mutual cache evictions are eliminated. To achieve a high WCET reduction, the involved call graph is based on worst-case call frequencies to focus on procedures on the WCEP.

WCET-Aware Trace Scheduling is a global instruction scheduling approach. Unlike local scheduling, WCC's trace scheduling operates on a trace which is constructed based on WCET data. By scheduling along this trace, maximal instruction-level parallelism is exposed on the WCEP.

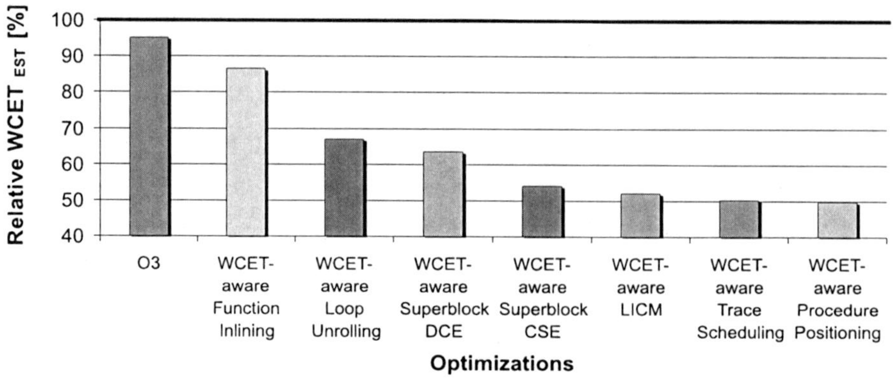

Fig. 8.1 Relative WCET estimates for entire optimization sequence

WCET-Aware Loop-Invariant Code Motion reduces the number of executed loop-invariant instructions by moving them outside the loop. Due to positive but also negative effects of this transformation on the WCET, the construction of an effective heuristic is not trivial. To automatically generate a heuristic for WCET reduction, machine learning techniques were exploited. To enhance the performance of the learned heuristics, the parameter settings of the involved machine learning algorithms were optimized via evolutionary algorithms.

Finally, to demonstrate the high performance of the WCC framework for an automatic WCET reduction, an overall evaluation of the developed optimization techniques is presented for the real-life benchmark *fir2dim* [ZVS+94]. Since the benchmark does not exhibit any potential for WCET-aware procedure cloning and unswitching, the optimizations were excluded here. The remaining optimizations were applied in the following order:

- WCET-aware Function Inlining
- WCET-aware Loop Unrolling
- WCET-aware Superblock DCE
- WCET-aware Superblock CSE
- WCET-aware Loop-Invariant Code Motion (LICM)
- WCET-aware Trace Scheduling
- WCET-aware Procedure Positioning

Figure 8.1 shows the relative WCETs achieved by this combination of the novel techniques. The baseline of 100% corresponds to the WCET of the non-optimized code. The first bar represents the relative WCET when the benchmark is compiled with WCC's highest optimization level *O3*. The remaining bars indicate the relative WCET when additional optimizations are applied on top of *O3*, i.e., the second bar represents the relative WCET for WCET-aware Function Inlining, the third bar corresponds to the relative WCET for the optimization sequence WCET-aware Function Inlining and Loop Unrolling etc.

As can be seen, standard ACET optimizations performed in *O3* achieve a WCET reduction of 4.8%. Applying additional WCET-aware optimizations leads to a continuous reduction of the WCET, i.e., the improvements achieved by each optimization add up if applied in combination. In total, the WCET of *fir2dim* could be reduced by 50.3%, outperforming *O3* by 45.5%. These results emphasize the fact that even more significant WCET savings can be achieved if the proposed WCET-aware optimization techniques are not applied separately but in concert.

Besides the achieved savings of the worst-case execution time using the presented optimization techniques in this book, the contributions of the presented techniques are:

- This work can be seen as a first systematic evaluation of the impact of compiler optimizations on the worst-case performance of embedded software.
- The achieved reductions of the WCET by the novel optimizations indicated that it is worthwhile to study well-known compiler optimizations in order to identify opportunities for trimming them for an explicit improvement of the program's WCET.
- The evaluation of the presented optimizations pointed out that WCET-aware optimizations may have a different impact on the WCET and ACET. Therefore, it could be concluded that also traditional ACET optimizations may not be suitable for an effective WCET reduction, emphasizing the need for techniques tailored towards the optimization of real-time systems.
- For the proposed optimization techniques, generic algorithms were presented. Thus, these techniques can be easily integrated into future real-time compilers.
- The proposed techniques help to cut down costs in two different ways. On the one hand, the proposed framework automates the software generation flow, thus shortens the real-time system design process. On the other hand, the reduction of the application's worst-case performance allows to reduce product costs since tailored hardware can be utilized.

8.2 Future Work

Despite the high WCET improvements achieved by the proposed techniques, there is always room for more research, striving for optimality of the program performance. This book ends with a brief outlook on the most relevant directions for future research, sorted by the individual chapters:

Extensions to WCC Compiler Framework   The WCC compiler framework has shown its capabilities for an effective reduction of the WCET of a single task. However, real-time systems often operate as multi-task systems where an operating system schedules the execution of each single task. To improve the worst-case behavior of the entire system, a promising research direction is the establishment of a communication between the WCET-aware compiler and the operating system. The resulting exchange of information could improve both the schedulability analysis and the tightness of WCET results due to additional data provided to the timing analysis. Another highly rewarding research direction is the extension of the WCC infrastructure towards multi-processors. This way, the software design flow for these systems could be automated and novel WCET-aware optimizations could be developed.

WCET-Aware Source Code Level Optimizations	This class of optimizations has shown to be highly effective for WCET reduction (cf. Chap. 4). Using back-annotation, assembly level information becomes available at the source code level, providing opportunities for novel optimization strategies. Therefore, it is a reasonable research direction to extend further traditional optimizations for a systematic improvement of the program's worst-case performance.

WCET-Aware Assembly Level Optimizations	The proposed WCET-aware global instruction scheduling (cf. Sect. 5.4) could be further evaluated for other parameter settings. For example, the impact of the maximally permitted length of a trace could be further investigated. Moreover, the trace scheduling could be extended towards a WCET-aware superblock scheduling to avoid the generation of compensation code. Another promising research direction is the development of source code and assembly level optimizations for multi-processor systems to reduce the WCET. For example, multi-processor scratchpad allocation is an interesting area which offers high potential for increased predictability of these complex systems.

Machine Learning Techniques in Compiler Design	The discussion in Chap. 6 has shown that machine learning can automate the generation of effective heuristics at both source code and assembly level. Research in this direction should be pursued for other optimizations, e.g., to generate WCET-aware heuristics for the optimizations register allocation or global instruction scheduling. Moreover, the model selection problem of machine learning—i.e., the choice of learning algorithms and their respective parameters—could be improved by the *feature selection* which finds promising features from the set of extracted features. As shown in [WM07], using an appropriate representation for training is beneficial for every learner since different learners show different preferences for the representation of features.

Multi-objective Optimizations	The search for Pareto optimal optimization sequences as presented in Chap. 7 can be extended in several directions. Up to now, the experiments were conducted for a single hardware configuration. It would be interesting to figure out how the optimization sequences change when architectural parameters were modified, such as I-cache capacity or the cache associativity. Moreover, further objectives that are critical for embedded/cyper-physical systems, e.g., energy consumption, could be involved in the search for Pareto optimal sequences. Also, Pareto fronts of more than two dimensions could be studied. Finally, approaches to speed-up the search process could be investigated. For this purpose, techniques known from the context of iterative compilation for a single objective, such as search space characterization [ACG+04], should be rethought and possibly adjusted to multi-objective optimization.

Appendix A
Abstract Interpretation

This appendix introduces the mathematical and fundamental theorems underlying abstract interpretation. The presentation is loosely based on [NNH99, CC77, Gus00]. The goal of the following discussion is to indicate why complexity problems may be encountered during the computation of concrete program semantics and to provide a possible solution to these problems. The solution presented in this work is based on abstract interpretation that reduces computational complexity by introducing an approximation of program semantics.

The goal of a program analysis is to extract information about the program behavior based on the computation of concrete program semantics.

A.1 Concrete Semantics

In a semantics-based program analysis, the program under analysis is described by *states*. Given a set *VAR* of all possible program variables and a set V of possible values which can be assigned to the program variables, a state is defined as:

Definition A.1 (State) A *state* σ is an assignment of values to variables such that $\sigma = \{var_1 \mapsto val_1, \ldots, var_n \mapsto val_n\}$, with $var \in$ VAR and $val \in V$. The set STATE of all variable bindings is defined as: STATE = VAR $\to V$.

A *program control point* θ corresponds to an edge $e \in E$ in a control flow graph $G = (V, E, i)$, referring to the next statement in $\mathcal{P}$ to be executed. Let PC be the set of program control points. A *configuration* is defined as:

Definition A.2 (Configuration) A *configuration* c of a program $\mathcal{P}$ is a pair $\langle \theta, \sigma \rangle$. The set CONF of configurations of the program is defined as CONF = PC $\times$ STATE.

The *transition function* $\longrightarrow: CONF \times CONF$ models a single execution step of the program based on the concrete semantics of the programming language of $\mathcal{P}$.

P. Lokuciejewski, P. Marwedel, *Worst-Case Execution Time Aware*
Compilation Techniques for Real-Time Systems, Embedded Systems,
DOI 10.1007/978-90-481-9929-7, © Springer Science+Business Media B.V. 2011

Let $\pi = (\theta_0, \ldots, \theta_t) \in V^*$ be a path through the control flow graph where $c \xrightarrow{*} c'$ denotes the *derivation sequence* $c \to \cdots \to c_k \to \cdots \to c'$ between configurations. The *path semantics* $\mathcal{PS}$ of $\mathcal{P}$ starting at the program control point θ_0 with the initial state σ_0 and terminating at the program control point θ_t is defined as a partial function from the initial state to a termination state σ_t:

Definition A.3 (Path semantics) $\mathcal{PS}[\![\pi]\!]\sigma_0 = \begin{cases} \sigma_t & \text{if } \langle \theta_0, \sigma_0 \rangle \xrightarrow{*} \langle \theta_t, \sigma_t \rangle \\ \bot & \text{otherwise} \end{cases}$

If the termination point is reached, the path semantics of the program is represented by the termination state σ_t, otherwise the path semantics is not defined.

The path semantics is computed for a single configuration. Program analysis should usually determine properties for sets of initial states. The *collecting semantics* $\mathcal{CS}$ of a program is an abstraction of the path semantics. Instead of collecting the history of computations in a predefined order, the collecting semantics is skipping information of execution order and collects all configurations that can be reached while executing a program. For the collecting semantics, the transition function must be lifted from individual configurations to sets of concrete states.

Let $\wp(CONF)$ be the power set of set $CONF$. The *set transition function* $\longrightarrow_\wp: \wp(CONF) \to \wp(CONF)$ denotes a set of configurations that can be reached in one transition step from a preceding configuration set. Note, that the set transition function is monotone by construction since it computes more precise information as output when more precise information is provided as input. The collecting semantics $\mathcal{CS} \in \wp(CONF)$ starting at a program control point θ_0 is obtained by applying the set transition function $\longrightarrow_\wp$ to an initial set C_0 of configurations and then forming a union of the computed sets.

Definition A.4 (Collecting semantics) The *collecting semantics* $\mathcal{CS}[\![\theta_0]\!]C_0$ for an initial program control point θ_0 and an initial configuration set C_0 is defined as:

$$\mathcal{CS}[\![\theta_0]\!]C_0 = \bigcup_{i \geq 0} C_0^i$$

$$C_0^0 = C_0$$

$$C_0^{i+1} = \longrightarrow_\wp C_0^i$$

with C_0^i being the set of configurations reached after i steps starting from C_0. Note that the domain used in collecting semantics is a *complete lattice* [NNH99]:

Definition A.5 (Complete lattice) A *complete lattice* $L = (L, \sqsubseteq, \sqcup, \sqcap, \bot, \top)$ is a partially ordered set $(L, \sqsubseteq)$ such that all subsets have a least upper bound as well as a greatest lower bound. Let $\sqcap L$ be the least upper bound of L and $\sqcup L$ be the greatest lower bound of L. Then, $\bot = \sqcup \emptyset = \sqcap L$ is the least element and $\top = \sqcap \emptyset = \sqcup L$ is the greatest element.

In the concrete domain, the complete lattice L for a superset of *CONF* is represented as $L = (\wp(\text{CONF}), \subseteq, \cup, \cap, \emptyset, \text{CONF})$. The partial order $\subseteq$ orders configurations with respect to their precision, i.e., if $A \subseteq B \in CONF$, then set A is more precise than set B and A contains fewer configurations.

As an alternative, program analyses often determine properties of the *sticky collecting semantics* $\mathcal{SCS}$ where each program point θ is associated with its possible set of states: $\mathcal{SCS} : PC \times \wp(STATE)$.

Definition A.6 (Sticky collecting semantics)

$$\mathcal{SCS}(\theta_i) = \{\sigma \mid \langle \theta_i, \sigma \rangle \in \mathcal{CS}\}$$

The sticky collecting semantics can be calculated by solving a set of recursive equations, so called *data flow equations*, that model the effect of all possible transitions to the states at a certain program point. A solution can be found by calculating the *least fixed-point* with methods such as the *Jacobi iteration*.

A common property of the collecting and the sticky collecting semantics is their precision since they determine all states encountered during iteration. However, in general they are not computable if the set of concrete initial states is of infinite size. For a bounded input domain found in practice, the semantics is still not computable due to feasibility reasons since testing of all inputs can not be performed in an acceptable time. In addition, it is in general desirable to force a termination of the analysis even if the program under analysis does not terminate. To overcome the first problem concerning feasibility, the concrete domain can be replaced by a (smaller, i.e., less precise) abstract domain. This makes the analysis feasible but usually also introduces imprecision. The termination of calculation can be achieved by a safe extrapolation of values during the calculation. Both techniques can be accomplished with the abstract interpretation which is discussed in the following.

A.2 Abstract Interpretation

Abstract interpretation is a general framework for a sound approximation of concrete semantics into abstract semantics. In the following, the abstraction of the collecting semantics is briefly discussed.

A.2.1 Abstract Semantics

The concrete domain D_{con}, which is represented by the lattice $(\wp(STATE), \subseteq)$, can be described by an abstract domain $D_{abs} = (D_{abs}, \sqsubseteq)$ with a partial order $\sqsubseteq$, i.e., $a \sqsubseteq b$ means that a denotes more precise analysis information than b. The abstract domain D_{abs} must be a complete lattice, i.e., for all subsets of D_{abs} least upper bounds exist that allow a safe and unique combination of analysis results.

Fig. A.1 Galois connection

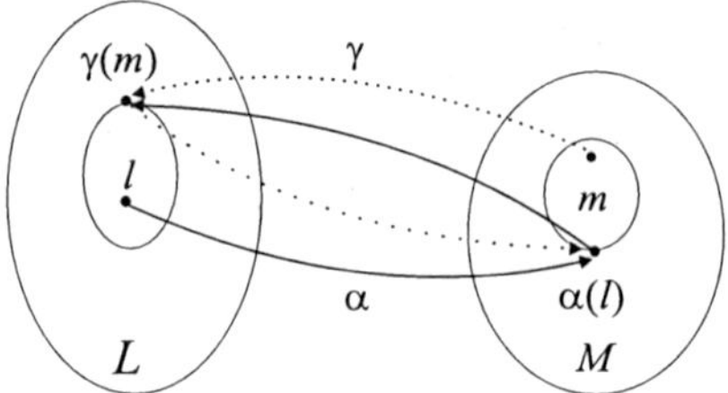

In order to relate the abstract and the concrete domain, the respective lattices are connected via a dual function. The monotone *abstraction function* $\alpha : D_{con} \to D_{abs}$ maps elements in the concrete domain to approximating elements while respecting the partial order. The monotone *concretization function* $\gamma : D_{abs} \to D_{con}$ maps each abstract state to a set of concrete states which it represents. Since γ is monotone, the partial order is respected, i.e., if $a \sqsubseteq b$ then $\gamma(a) \subseteq \gamma(b)$. The abstraction function α and the concretization function γ form a *Galois connection*:

Definition A.7 (Galois connection) Let $(L, \leq)$ and $(M, \sqsubseteq)$ be partially ordered sets and $\alpha : L \to M$, $\gamma : M \to L$. $L \leftrightarrows M$ is called a *Galois connection* if α and γ are monotone functions and

$$\forall l \in L \wedge \forall m \in M : l \leq \gamma(\alpha(l)) \wedge \alpha(\gamma(m)) \sqsubseteq m.$$

Figure A.1 illustrates this definition.

The first condition introduces imprecision when the abstraction and concretization function is applied, however it guarantees correctness. Since M is smaller than L, the loss of precision can usually be not avoided. The second condition ensures that α determines precise approximations of elements of the concrete domain. However, the condition also introduces surjection since two elements of the abstract domain describe the same element of the concrete domain. To avoid this undesired situation, the condition is strengthened to $\alpha(\gamma(m)) = m$. These stronger requirements for α and γ are called the *Galois insertion*.

A.2.2 Abstract Domain

The choice for a suitable abstract domain used by abstract interpretation depends on the requirements for the analysis and is often a trade-off between complexity and precision. Examples are the congruence domain [Gra89] or the octogan domain [Min06]. WCC's loop analyzer operates on the interval domain [CC77] which provides a good trade-off for the considered problem. Within the interval domain, possible values of a variable are approximated by an interval i:

$$i = \{\bot\} \cup \{[z_1, z_2] \mid z_1 \leq z_2\}$$

for $z_1 \in \mathbb{Z} \cup \{-\infty\}$, $z_2 \in \mathbb{Z} \cup \{\infty\}$ and $\bot$ representing an empty interval. As shown in [CC77], the interval domain represents a complete lattice for which the operators $\sqcap$, $\sqcup$, $\sqsubseteq$, $\top$, and $\bot$ are defined.

The interval domain allows to specify the *abstract state* σ_{abs} on the interval domain analogous to the concrete states:

Definition A.8 (Abstract state) An *abstract state* σ_{abs} is an assignment of intervals to variables:

$$STATE_{abs} = (VAR \rightarrow INTERVAL_\bot) \cup \{\bot_\sigma\} \cup \{\top_\sigma\}$$

with *VAR* the set of program variables, *INTERVAL* the set of all intervals including $\bot$ (empty interval) and $\top$ (any concrete values). In addition, $\bot_\sigma$ indicates an infeasible state, while $\sigma_{abs} = \top_\sigma$ represents any concrete state.

Using the definition of the abstract states, the abstract configuration is defined as follows:

Definition A.9 (Abstract configuration) An *abstract configuration* c_{abs} of a program $\mathcal{P}$ is a pair $\langle \theta, \sigma_{abs} \rangle$. The set $CONF_{abs}$ of configurations of the program is defined as $CONF_{abs} = PC \times STATE_{abs}$.

The construction of an abstract interpretation requires the specification of monotone *abstract set transition functions* which are the counterpart of concrete transitions. The abstract function is defined as: $\xrightarrow{abs}_{\wp} : \wp(CONF_{abs}) \rightarrow \wp(CONF_{abs})$.

The abstract set transition functions specifies how a set of abstract configurations is computed in one transition step based on a previous set of configurations. $\xrightarrow{abs}_{\wp}$ is defined by abstract operators of a particular abstract domain which are based on the semantics of a given programming language. For example, the abstract addition operator $\hat{+}$ carried out on the interval domain is defined as:

$$A\hat{+}B = [\inf(A) + \inf(B), \sup(A) + \sup(B)]$$

where inf and sup represent the infimum and supremum of the intervals A and B, respectively. For a complete list of abstract operators defined on the interval domain, the interested reader is referred to [Cor08, Gus00].

A.2.3 Calculation of Abstract Semantics

Analogous to the collecting semantics, the abstract semantics can be calculated as the fixed-point of a recursive set of abstract data flow equations. The equations resemble those of the concrete semantics with the only difference that concrete configurations, states, and transition functions are substituted by the respective abstract versions.

In some cases, the fixed-point iteration will not terminate since the data flow equations do not stabilize [GEL05]. This behavior can often be observed at loop headers of the CFG where the data flow equations express join points between the immediate loop predecessor and the loop back-edge. However, as mentioned previously, abstract interpretation must be constructed such that termination of the analysis is guaranteed even for non-terminating programs. Another problem might be a high analysis time which is often encountered for loops with large iteration counts. The reason is the iterative nature of the classical abstract interpretation relying on solving of data flow equations. To cope with both problems, the binary *widening operator* $\triangledown$ [CC92] may be inserted into the abstract data flow equations.

Definition A.10 (Widening operator) The *widening operator* $\triangledown \in L \times L \to L$ for the lattice of intervals $L = \{\bot\} \cup \{[l, u] \mid l \in \mathbb{Z} \cup \{-\infty\}\} \wedge u \in \mathbb{Z} \cup \{+\infty\}\} \wedge l \leq u\}$ is defined as:

$$\bot \triangledown X = X$$

$$X \triangledown \bot = X$$

$$[l_0, u_0] \triangledown [l_1, u_1] = [\text{if } l_1 < l_0 \text{ then } -\infty \text{ else } l_0,$$

$$\text{if } u_1 > u_0 \text{ then } +\infty \text{ else } u_0]$$

The widening operator extrapolates unstable bounds to infinity. Its application may rely on user specifications, e.g., after how many loop iterations without finding a fixed-point widening should be performed. The main drawback of the operator is the introduced loss of precision that possibly results in unbounded variable assignments that have to be propagated through the data flow equations. Sometimes, the results can be improved by the application of the *narrowing operator* which tries to improve infinite bounds. However, its application does often not suffice or is even impossible. Another problem of the widening operator is that there is no generic way to find suitable implementations of the widening/narrowing operators for every type of computations [Gus00]. Therefore, these operators should be used with caution.

Appendix B
Transformation of Conditions

This appendix gives an example for the computation of polytopes that is used during WCC's static loop analysis (cf. Sect. 3.6.4.1).

Given the condition:

$$if\ (\overline{(2*i+2 \leq j\ \text{NAND}\ j > 15)})$$

The goal is to transform the condition into the normalized form from Definition 3.7 (cf. p. 48).

The first step is a transformation of the conditions into an equivalent form of type $if\ (C_1 \oplus C_2 \oplus \cdots \oplus C_n)$ where C_x are conditions. The set of operators $\oplus$ consists of the elements $\{\wedge, \vee\}$. In addition, each expression can be negated by the *NOT* operator ($\overline{x}$). With the means of *propositional calculus* and *De Morgan's laws*, it is possible to transform each condition in this equivalent form.

In the given example condition, the *NAND* operator does not belong to the set $\oplus$ of supported operators and must thus be eliminated by the transformation:

$$x\ \text{NAND}\ y = \overline{x \wedge y}$$

Applying this rule to the example, the transformed condition is obtained:

$$if\ (\overline{(\overline{2*i+2 \leq j \wedge j > 15})})$$

Removing the double negation leads to

$$if\ (2*i+2 \leq j \wedge j > 15)$$

This condition corresponds to the form $if\ (C_1 \oplus C_2 \oplus \cdots \oplus C_n)$, thus finishing the first transformation phase.

P. Lokuciejewski, P. Marwedel, *Worst-Case Execution Time Aware Compilation Techniques for Real-Time Systems*, Embedded Systems, DOI 10.1007/978-90-481-9929-7, © Springer Science+Business Media B.V. 2011

In a second step, each condition C_i must have the normalized form $C_x = \sum_{l=1}^{N}(c_l * i_l) + c \geq 0$ for constant values c_l, c with c_l, $c \in \mathbb{Z}$ and i_l representing variables within the conditions. Each affine condition using the comparators $\oplus \in \{<, \leq, >, \geq, =\}$, can be transformed into this normalized form [FM03]. Normalizing the example condition results in

$$if\ (-2 * i + j - 2 \geq 0 \wedge j - 16 \geq 0)$$

References

Abs10.   AbsInt Angewandte Informatik GmbH, Worst-Case Execution Time Analyzer aiT for TriCore. http://www.absint.com/ait, March 2010

ABC+06.   F. Agakov, E. Bonilla, J. Cavazos et al., Using machine learning to focus iterative optimization, in *Proceedings of the 4th Annual IEEE/ACM International Symposium on Code Generation and Optimization (CGO)*, New York, USA, March 2006, pp. 295–305

ASU86.   A.V. Aho, R. Sethi, J.D. Ullman, *Compilers: Principles, Techniques, and Tools* (Addison-Wesley/Longman, Boston, 1986)

ACG+04.   L. Almagor, K.D. Cooper, A. Grosul et al., Finding Effective Compilation Sequences, in *Proceedings of the ACM SIGPLAN/SIGBED Conference on Languages, Compilers, and Tools for Embedded Systems (LCTES)*, Washington, USA, June 2004, pp. 231–239

App97.   A.W. Appel, *Modern Compiler Implementation in C* (Cambridge University Press, New York, 1997)

ARM01.   *ARM7TDMI-S (Revision 4), Technical Reference Manual*. ARM Limited, September 2001

BGS94.   D.F. Bacon, S.L. Graham, O.J. Sharp, Compiler transformations for high-performance computing. ACM Comput. Surv. **26**(4), 345–420 (1994)

BL99.   S. Bashford, R. Leupers, Phase-coupled mapping of data flow graphs to irregular data paths. Des. Autom. Embed. Syst. **4**(2), 1–50 (1999)

BH93.   S. Bates, S. Horwitz, Incremental program testing using program dependence graphs, in *Proceedings of the 20th ACM SIGPLAN-SIGACT Symposium on Principles of Programming Languages (POPL)*, Charleston, USA, January 1993, pp. 384–396

Bis08.   C.M. Bishop, *Pattern Recognition and Machine Learning* (Springer, Berlin, 2008)

BLTZ03.   S. Bleuler, M. Laumanns, L. Thiele, E. Zitzler, PISA—a platform and programming language independent interface for search algorithms, in *Proceedings of 2nd International Conference on Evolutionary Multi-Criterion Optimization (EMO)*, Faro, Portugal, April 2003, pp. 494–508

Bör96.   H. Börjesson, Incorporating worst case execution time in a commercial C-compiler, Master's thesis, Uppsala University, January 1996

BGCSV09.   P. Brazdil, C. Giraud-Carrier, C. Soares, R. Vilalta, *Metalearning: Applications to Data Mining* (Springer, Berlin, 2009)

Bre01.   L. Breiman, Random forests. Mach. Learn. **45**(1), 5–32 (2001)

Bri92.   P. Briggs, Register allocation via graph coloring, PhD thesis, Rice University, Houston, USA, 1992

BEGL05.   S. Byhlin, A. Ermedahl, J. Gustafsson, B. Lisper, Applying static WCET analysis to automotive communication software, in *Proceedings of the 17th Euromicro Conference of Real-Time Systems (ECRTS)*, Palma de Mallorca, Spain, July 2005, pp. 249–258

P. Lokuciejewski, P. Marwedel, *Worst-Case Execution Time Aware*   243
Compilation Techniques for Real-Time Systems, Embedded Systems,
DOI 10.1007/978-90-481-9929-7, © Springer Science+Business Media B.V. 2011

CGJ+97.   B. Calder, D. Grunwald, M. Jones et al., Evidence-based static branch prediction using machine learning. ACM Trans. Program. Lang. Syst. **19**(1), 188–222 (1997)

CPI+05.   A.M. Campoy, I. Puaut, A.P. Ivars et al., Cache contents selection for statically-locked instruction caches: an algorithm comparison, in *Proceedings of the 17th Euromicro Conference on Real-Time Systems (ECRTS)*, Palma de Mallorca, Spain, July 2005, pp. 49–56

CK94.   S. Carr, K. Kennedy, Improving the ratio of memory operations to floating-point operations in loops. ACM Trans. Program. Lang. Syst. **16**(6), 1768–1810 (1994)

CM04.   J. Cavazos, J.E.B. Moss, Inducing heuristics to decide whether to schedule. SIGPLAN Not. **39**(6), 183–194 (2004)

CO06.   J. Cavazos, M. O'Boyle, Method-specific dynamic compilation using logistic regression, in *Proceedings of the 21st Annual ACM SIGPLAN Conference on Object-oriented Programming Systems, Languages, and Applications (OOPSLA)*, Portland, USA, October 2006, pp. 229–240

CO05.   J. Cavazos, M. O'Boyle, Automatic tuning of inlining heuristics, in *Proceedings of the ACM/IEEE Conference on Supercomputing (SC)*, Seattle, USA, November 2005, pp. 14–25

CFA+07.   J. Cavazos, G. Fursin, F. Agakov et al., Rapidly selecting good compiler optimizations using performance counters, in *Proceedings of the International Symposium on Code Generation and Optimization (CGO)*, San Jose, USA, March 2007, pp. 185–197

CH88.   P.P. Chang, W.W. Hwu, Trace selection for compiling large C application programs to microcode, in *Proceedings of the 21st Annual Workshop on Microprogramming and Microarchitecture (MICRO)*, San Diego, USA, November 1988, pp. 21–29

CMH91.   P.P. Chang, S.A. Mahlke, W.W. Hwu, Using profile information to assist classic code optimizations. Softw. Pract. Exp. **21**(12), 1301–1321 (1991)

CMW+92.   W. Chen, S. Mahlke, N. Warter et al., Using profile information to assist advanced compiler optimization and scheduling. Adv. Lang. Compil. Parallel Process. **757**, 31–48 (1992)

Chv83.   V. Chvátal, *Linear Programming* (Freeman, New York, 1983)

CL00.   R. Cohn, P.G. Lowney, Design and analysis of profile-based optimization in Compaq's compilation tools for alpha. J. Instr. Level Parallelism **2**, 1–25 (2000)

CB02.   A. Colin, G. Bernat, Scope-Tree: a program representation for symbolic worst-case execution time analysis, in *Proceedings of the 14th Euromicro Conference on Real-Time Systems (ECRTS)*, July 2002, pp. 50–59

CP01.   A. Colin, I. Puaut, A modular & retargetable framework for tree-based WCET analysis, in *Proceedings of the 13th Euromicro Conference on Real-Time Systems (ECRTS)*, Delft, The Netherlands, June 2001, pp. 37–44

CNO+87.   R.P. Colwell, R.P. Nix, J.J. O'Donnell et al., A VLIW architecture for a trace scheduling compiler. ACM SIGPLAN Not. **22**(10), 180–192 (1987)

Con71.   W.J. Conover, *Practical Nonparametric Statistics* (Wiley, New York, 1971)

CT04.   K.D. Cooper, L. Torczon, *Engineering A Compiler* (Morgan Kaufmann, San Francisco, 2004)

CHT91.   K.D. Cooper, M.W. Hall, L. Torczon, An experiment with inline substitution. Softw. Pract. Exp. **21**(6), 581–601 (1991)

CHK93.   K.D. Cooper, M.W. Hall, K. Kennedy, A methodology for procedure cloning. Comput. Lang. **19**(2), 105–117 (1993)

CSS99.   K.D. Cooper, P.J. Schielke, D. Subramanian, Optimizing for reduced code space using genetic algorithms. ACM SIGPLAN Not. **34**(7), 1–9 (1999)

CST02.   K.D. Cooper, D. Subramanian, L. Torczon, Adaptive optimizing compilers for the 21st century. J. Supercomput. **23**(1), 7–22 (2002)

Cor08.   D. Cordes, Loop analysis for a WCET-optimizing compiler based on abstract interpretation and polylib (in German), Diploma thesis, TU Dortmund University, April 2008

CC77. P. Cousot, R. Cousot, Abstract interpretation: a unified lattice model for static analysis of programs by construction or approximation of fixpoints, in *Proceedings of the 4th ACM SIGACT-SIGPLAN Symposium on Principles of Programming Languages (POPL)*, Los Angeles, USA, January 1977, pp. 238–252

CC92. P. Cousot, R. Cousot, Comparing the Galois connection and widening/narrowing approaches to abstract interpretation, in *Proceedings of the 4th International Symposium on Programming Language Implementation and Logic Programming (PLILP)*, Leuven, Belgium, August 1992, pp. 269–295

CM07. C. Cullmann, F. Martin, Data-flow based detection of loop bounds, in *Proceedings of the 7th International Workshop on Worst-Case Execution Time (WCET)*, Pisa, Italy, July 2007, pp. 57–62

DH89. J.W. Davidson, A.M. Holler, Subprogram inlining: a study of its effects on program execution time. Technical report, University of Virginia, Charlottesville, USA, 1989

DJ01. J.W. Davidson, S. Jinturkar, An aggressive approach to loop unrolling, Technical report, University of Virginia, Charlottesville, USA, 2001

DAPM00. K. Deb, S. Agrawal, A. Pratap, T. Meyarivan, A fast elitist non-dominated sorting genetic algorithm for multi-objective optimisation: NSGA-II, in *Proceedings of the 6th International Conference on Parallel Problem Solving from Nature (PPSN)*, Paris, France, September 2000, pp. 849–858

DP07. J.F. Deverge, I. Puaut, WCET-directed dynamic scratchpad memory allocation of data, in *Proceedings of the 19th Euromicro Conference on Real-Time Systems (ECRTS)*, Pisa, Italy, July 2007, pp. 179–190

EF94. D. Elsner, J. Fenlason, *Using as—The GNU Assembler*. Free Software Foundation (1994)

EE00. J. Engblom, A. Ermedahl, Modeling complex flows for worst-case execution time analysis, in *Proceedings of the 21st IEEE Real-Time Systems Symposium (RTSS)*, Orlando, USA, November 2000, pp. 163–174

Erm03. A. Ermedahl, A modular tool architecture for worst-case execution time analysis, PhD thesis, Uppsala University, 2003

EG97. A. Ermedahl, J. Gustafsson, Deriving annotations for tight calculation of execution time, in *Proceedings of the 3rd International Euro-Par Conference on Parallel Processing (Euro-Par)*, Passau, Germany, August 1997, pp. 1298–1307

ESG+07. A. Ermedahl, C. Sandberg, J. Gustafsson et al., Loop bound analysis based on a combination of program slicing, abstract interpretation, and invariant analysis, in *Proceedings of the 7th International Workshop on Worst-Case Execution Time Analysis (WCET)*, Pisa, Italy, July 2007, pp. 63–68

EH94. A. Erosa, L.J. Hendren, Taming control flow: a structured approach to eliminating goto statements, in *Proceedings of IEEE International Conference on Computer Languages (ICCL)*, Toulouse, France, May 1994, pp. 229–240

ETA10. ETAS Group, ASCET Software Products, http://www.etas.de, March 2010

Fal09. H. Falk, WCET-aware register allocation based on graph coloring, in *Proceedings of the 46th Design Automation Conference (DAC)*, San Francisco, USA, July 2009, pp. 726–731

FK09. H. Falk, J.C. Kleinsorge, Optimal static WCET-aware scratchpad allocation of program code, in *Proceedings of the 46th Design Automation Conference (DAC)*, San Francisco, USA, July 2009, pp. 732–737

FM03. H. Falk, P. Marwedel, Control flow driven splitting of loop nests at the source code level, in *Proceedings of the Conference on Design, Automation and Test in Europe (DATE)*, Munich, Germany, March 2003, pp. 410–415

FS06. H. Falk, M. Schwarzer, Loop nest splitting for WCET-optimization and predictability improvement, in *Proceedings of the 2006 IEEE/ACM/IFIP Workshop on Embedded Systems for Real Time Multimedia (ESTIMedia)*, Seoul, Korea, October 2006, pp. 115–120

FLT06a. H. Falk, P. Lokuciejewski, H. Theiling, Design of a WCET-aware C compiler, in *Proceedings of the 2006 IEEE/ACM/IFIP Workshop on Embedded Systems for Real Time Multimedia (ESTIMedia)*, Seoul, Korea, October 2006, pp. 121–126

FLT06b. H. Falk, P. Lokuciejewski, H. Theiling, Design of a WCET-aware C compiler, in *Proceedings of the 6th International Workshop on Worst-Case Execution Time Analysis (WCET)*, Dresden, Germany, June 2006

FPT07. H. Falk, S. Plazar, H. Theiling, Compile-time decided instruction cache locking using worst-case execution paths, in *Proceedings of the 5th IEEE/ACM International Conference on Hardware/software Codesign and System Synthesis (CODES+ISSS)*, Salzburg, Austria, September 2007, pp. 143–148

FBH+06. H. Fennel, S. Bunzel, H. Heinecke et al., Achievements and exploitation of the AUTOSAR development partnership, in *SAE Convergence*, October 2006

FW99. C. Ferdinand, R. Wilhelm, Fast and efficient cache behavior prediction for real-time systems. Real-Time Syst. **17**, 131–181 (1999)

FHL+01. C. Ferdinand, R. Heckmann, M. Langenbach et al., Reliable and precise WCET determination for a real-life processor, in *Proceedings of the 1st International Workshop on Embedded Software (EMSOFT)*, Tahoe City, USA, October 2001, pp. 496–485

FOW87. J. Ferrante, K.J. Ottenstein, J.D. Warren, The program dependence graph and its use in optimization. ACM Trans. Program. Lang. Syst. **9**(3), 319–349 (1987)

Fis81. J.A. Fisher, Trace scheduling: a technique for global microcode compaction. IEEE Trans. Comput. **30**(7), 478–490 (1981)

FHP92. C.W. Fraser, R.R. Henry, T.A. Proebsting, BURG: fast optimal instruction selection and tree parsing. SIGPLAN Not. **27**(4), 68–76 (1992)

Fur05. G. Fursin, A heuristic search algorithm based on unified transformation framework, in *Proceedings of the 2005 International Conference on Parallel Processing Workshops (ICPPW)*, Oslo, Norway, June 2005, pp. 137–144

FMP+07. G. Fursin, C. Miranda, S. Pop et al., Practical run-time adaptation with procedure cloning to enable continuous collective compilation, in *Proceedings of the GCC Developers' Summit*, Ottawa, Canada, July 2007

GP07. R. Garside, J.F. Pighetti, Integrating modular avionics: a new role emerges, in *Proceedings of the 26th Digital Avionics Systems Conference (DASC)*, Dallas, USA, 2007, pp. 17–22

GCC10. GCC, GNU Compiler Collection, http://gcc.gnu.org/, March 2010

GH98. R. Ghiya, L.J. Hendren, Putting pointer analysis to work, in *Proceedings of the 25th ACM SIGPLAN-SIGACT Symposium on Principles of Programming Languages (POPL)*, San Diego, USA, January 1998, pp. 121–133

GMW81. R. Giegerich, U. Möncke, R. Wilhelm, Invariance of approximate semantics with respect to program transformations, in *GI - 11. Jahrestagung in Verbindung mit Third Conference of the European Co-operation in Informatics (ECI)*, Munich, Germany, October 1981, pp. 1–10

GW96. D.W. Goodwin, K.D. Wilken, Optimal and near-optimal global register allocation using 0–1 integer programming. Softw. Pract. Exp. **26**(8), 929–965 (1996)

Gra69. R.L. Graham, Bounds on multiprocessing timing anomalies. SIAM J. Appl. Math. **17**, 416–429 (1969)

Gra89. P. Granger, Static analysis of arithmetical congruences. Int. J. Comput. Math. **30**(30), 165–190 (1989)

GSC07. Y. Guo, D. Subramanian, K.D. Cooper, An effective local search algorithm for an adaptive compiler, in *Proceedings of the 1st Workshop on Statistical and Machine Learning Approaches Applied to Architectures and Compilation (SMART)*, Ghent, Belgium, January 2007, pp. 7–11

Gus00. J. Gustafsson, Analyzing execution-time of object-oriented programs using abstract interpretation, PhD thesis, Uppsala University, 2000

GEL05. J. Gustafsson, A. Ermedahl, B. Lisper, Towards a flow analysis for embedded system C programs, in *Proceedings of the 10th IEEE International Workshop on Object-*

Oriented Real-Time Dependable Systems (WORDS), Sedona, USA, February 2005, pp. 287–300

GES+06. J. Gustafsson, A. Ermedahl, C. Sandberg et al., Automatic derivation of loop bounds and infeasible paths for WCET analysis using abstract execution, in *The 27th IEEE Real-Time Systems Symposium (RTSS 2006)*, Rio de Janeiro, Brazil, December 2006, pp. 57–66

GRE+01. M. Guthaus, J. Ringenberg, D. Ernst et al., MiBench: a free, commercially representative embedded benchmark suite, in *Proceedings of the 4th IEEE International Workshop on Workload Characteristics (WWC)*, Austin, USA, December 2001, pp. 3–14

HKW05. M. Haneda, P. Knijnenburg, H. Wijshoff, Automatic selection of compiler options using non-parametric inferential statistics, in *Proceedings of the 14th International Conference on Parallel Architectures and Compilation Techniques (PACT)*, Saint Louis, USA, September 2005, pp. 123–132

HKK04. B. Hardung, T. Kölzow, A. Krüger, Reuse of software in distributed embedded automotive systems, in *Proceedings of the 4th ACM International Conference on Embedded Software (EMSOFT)*, Pisa, Italy, September 2004, pp. 203–210

HSR+98. C. Healy, M. Sjödin, V. Rustagi et al., Bounding loop iterations for timing analysis, in *Proceedings of the 4th IEEE Real-Time Technology and Applications Symposium (RTAS)*, Denver, USA, June 1998, pp. 12–21

HLT+03. R. Heckmann, M. Langenbach, S. Thesing et al., The influence of processor architecture on the design and the results of WCET tools. Proc. IEEE **91**(7), 1038–1054 (2003)

HP03. J.L. Hennessy, D.A. Patterson, *Computer Architecture: A Quantitative Approach* (Morgan Kaufmann, San Francisco, 2003)

HBK+01. K. Heydemann, F. Bodin, P. Knijnenburg et al., UFC: a global trade-off strategy for loop unrolling for VLIW architecture, in *Proceedings of the 10th Workshop on Compilers for Parallel Computers (CPC)*, Amsterdam, The Netherlands, January 2001, pp. 59–70

Hig10. HighTec EDV-Systeme GmbH. TriCore PXROS-HR Development Platform, http://www.hightec-rt.com, March 2010

HS89. M. Hill, A. Smith, Evaluating associativity in CPU caches. IEEE Trans. Comput. **38**(12), 1612–1630 (1989)

Hol92. J.H. Holland, *Adaptation in Natural and Artificial Systems* (MIT Press, Cambridge, 1992)

HGB+08. N. Holsti, J. Gustafsson, G. Bernat, C. Ballabriga, A. Bonenfant, R. Bourgade, H. Cassi, D. Cordes, A. Kadlec, R. Kirner, J. Knoop, P. Lokuciejewski, N. Merriam, M. de Michiel, A. Prantl, B. Rieder, C. Rochange, M. Sainrat, P. Schordan, WCET Tool Challenge 2008: Report, in *Proceedings of the 8th International Workshop on Worst-Case Execution Time Analysis (WCET)*, Prague, Czech Republic, July 2008

HLS00. N. Holsti, T. Langbacka, S. Saarinen, Using a worst-case execution time tool for real-time verification of the Debie software, in *Proceedings of the Conference on Data Systems in Aerospace (DASIA)*, Montreal, Canada, May 2000, p. 307

HRB88. S. Horwitz, T. Reps, D. Binkley, Interprocedural slicing using dependence graphs, in *Proceedings of the ACM SIGPLAN 1988 Conference on Programming Language Design and Implementation (PLDI)*, Atlanta, USA, June 1988, pp. 35–46

HE08. K. Hoste, L. Eeckhout, COLE: Compiler Optimization Level Exploration, in *Proceedings of the 6th Annual IEEE/ACM International Symposium on Code Generation and Optimization (CGO)*, Boston, USA, April 2008, pp. 165–174

HXV+05. W.L. Hung, Y. Xie, N. Vijaykrishnan et al., Thermal-aware task allocation and scheduling for embedded systems, in *Proceedings of the Conference on Design, Automation and Test in Europe (DATE)*, Munich, Germany, March 2005, pp. 898–899

HC89. W.W. Hwu, P.P. Chang, Achieving high instruction cache performance with an optimizing compiler. ACM SIGARCH Comput. Archit. News **17**(3), 242–251 (1989)

HMC+93. W.W. Hwu, S.A. Mahlke, W.Y. Chen et al., The superblock: an effective technique for VLIW and superscalar compilation. J. Supercomput. **7**, 229–248 (1993)

Inf10a.   Informatik Centrum Dortmund. ICD-C Compiler framework, http://www.icd.de/es/
           icd-c, March 2010
Inf10b.   Informatik Centrum Dortmund. ICD low level intermediate representation backend
           infrastructure (LLIR)—developer manual. Informatik Centrum Dortmund, March
           2010
ITK+03.   K. Ishizaki, M. Takeuchi, K. Kawachiya et al., Effectiveness of cross-platform op-
           timizations for a Java just-in-time compiler. ACM SIGPLAN Not. **38**(11), 187–204
           (2003)
KC02.   M. Kandemir, A. Choudhary, Compiler-directed scratch pad memory hierarchy design
           and management, in *Proceedings of the 39th annual Design Automation Conference
           (DAC)*, New Orleans, USA, June 2002, pp. 628–633
Kel09.   T. Kelter, Superblock-based high-level WCET optimizations, Diploma thesis, TU
           Dortmund University, September 2009 (in German)
KH06.   R. Kidd, W.W. Hwu, Abstract improved superblock optimization in GCC, in *GCC
           Summit* (2006)
Kir02.   R. Kirner, The programming language wcetC, Technical report, Technische Univer-
           sität Wien, Institut für Technische Informatik, 2002
Kir03.   R. Kirner, Extending optimising compilation to support worst-case execution time
           analysis, PhD thesis, Vienna University of Technology, 2003
Kir06.   M. Kirner, Automatic loop bound analysis of programs written in C, Master's thesis,
           Technische Universität Wien, December 2006
Kle08.   J. Kleinsorge, WCET-centric code allocation for scratchpad memories, Diploma the-
           sis, TU Dortmund University, September 2008
KTZ05.   J. Knowles, L. Thiele, E. Zitzler, A tutorial on the performance assessment of stochas-
           tic multiobjective optimizers, in *Proceedings of 3rd International Conference on Evo-
           lutionary Multi-Criterion Optimization (EMO)*, Guanajuato, Mexico, March 2005
KCS06.   A. Konak, D.W. Coit, A.E. Smith, Multi-objective optimization using genetic algo-
           rithms: a tutorial. Reliab. Eng. Syst. Saf. **91**(9), 992–1007 (2006)
KKF97.   A. Koseki, H. Komastu, Y. Fukazawa, A method for estimating optimal unrolling
           times for nested loops, in *Proceedings of the 1997 International Symposium on Par-
           allel Architectures, Algorithms and Networks (ISPAN)*, Washington, USA, December
           1997, pp. 376–382
KHW+05.   P.A. Kulkarni, S.R. Hines, D. Whalley et al., Fast and efficient searches for effective
           optimization-phase sequences. Trans. Archit. Code Optim. **2**(2), 165–198 (2005)
KBT+04.   S. Künzli, S. Bleuler, L. Thiele et al., A computer engineering benchmark application
           for multiobjective optimizers, in *Application of Multi-Objective Evolutionary Algo-
           rithms* (World Scientific, Singapore, 2004), pp. 269–294
LR91.   W. Landi, B.G. Ryder, Pointer-induced aliasing: a problem taxonomy, in *Proceed-
           ings of the 18th ACM SIGPLAN-SIGACT Symposium on Principles of Programming
           Languages (POPL)*, Orlando, USA, January 1991, pp. 93–103
LTZ+02.   M. Laumanns, L. Thiele, E. Zitzler et al., Running time analysis of multi-objective
           evolutionary algorithms on a simple discrete optimization problem, in *Proceedings of
           the 7th International Conference on Parallel Problem Solving from Nature (PPSN)*,
           Granada, Spain, September 2002, pp. 44–53
LH95.   D.M. Lavery, W.W. Hwu, Unrolling-based optimizations for modulo scheduling, in
           *Proceedings of the 28th Annual International Symposium on Microarchitecture (MI-
           CRO)*, Ann Arbor, USA, November 1995, pp. 327–337
LBO09.   H. Leather, E. Bonilla, M. O'Boyle, Automatic feature generation for machine learn-
           ing based optimizing compilation, in *Proceedings of the International Symposium on
           Code Generation and Optimization (CGO)*, Seattle, USA, March 2009, pp. 81–91
LOW09.   H. Leather, M. O'Boyle, B. Worton, Raced profiles: efficient selection of competing
           compiler optimizations, in *Proceedings of the ACM SIGPLAN/SIGBED Conference
           on Languages, Compilers, and Tools for Embedded Systems (LCTES)*, Dublin, Ire-
           land, June 2009, pp. 50–59

LW94.   A.R. Lebeck, D.A. Wood, Cache profiling and the SPEC benchmarks: a case study. IEEE Comput. **27**(10), 16–26 (1994)

Lee05.   E.A. Lee, Absolutely positively on time: what would it take? Computer **38**(7), 85–87 (2005)

Lee07.   E.A. Lee, Computing foundations and practice for cyber-physical systems: a preliminary report, Technical Report UCB/EECS-2007-72, University of California, Berkeley, May 2007

LLPM04.   S. Lee, J. Lee, C.Y. Park, S.L. Min, A flexible tradeoff between code size and WCET using a dual instruction set processor, in *Proceedings of the 8th International Workshop on Software & Compilers for Embedded Systems (SCOPES)*, Amsterdam, The Netherlands, September 2004, pp. 244–258

LPMS97.   C. Lee, M. Potkonjak, W.H. Mangione-Smith, MediaBench: a tool for evaluating and synthesizing multimedia and communications systems, in *Proceedings of the 30th Annual International Symposium on Microarchitecture (MICRO)*, Research Triangle Park, USA, December 1997, pp. 330–335

LT79.   T. Lengauer, R.E. Tarjan, A fast algorithm for finding dominators in a flowgraph. ACM Trans. Program. Lang. Syst. **1**(1), 121–141 (1979)

Leu00.   R. Leupers, Code selection for media processors with SIMD instructions, in *Proceedings of the Conference on Design, Automation and Test in Europe (DATE)*, Paris, France, March 2000, pp. 4–8

LMW95.   Y.T.S. Li, S. Malik, A. Wolfe, Efficient microarchitecture modeling and path analysis for real-time software, in *Proceedings of the 16th IEEE Real-Time Systems Symposium (RTSS)*, Pisa, Italy, May 1995, pp. 298–307

LLM+07.   X. Li, Y. Liang, T. Mitra et al., Chronos: a timing analyzer for embedded software. Sci. Comput. Program. **69**(1–3), 56–67 (2007)

LJC+10.   Y. Liang, L. Ju, S. Chakraborty et al., Cache-aware optimization of BAN applications. ACM Trans. Des. Automat. Electron. Syst. (2010)

LPJ96.   C. Liem, P. Paulin, A. Jerraya, Address calculation for retargetable compilation and exploration of instruction-set architectures, in *Proceedings of the 33rd annual Design Automation Conference (DAC)*, Las Vegas, USA, June 1996, pp. 597–600

LFMT08.   P. Lokuciejewsi, H. Falk, P. Marwedel, H. Theiling, WCET-driven, code-size critical procedure cloning, in *Proceedings of the 11th International Workshop on Software & Compilers for Embedded Systems (SCOPES)*, Munich, Germany, March 2008, pp. 21–30

Lok05.   P. Lokuciejewski, Design and realization of concepts for WCET compiler optimization, Diploma thesis, TU Dortmund University, December 2005

LM09.   P. Lokuciejewski, P. Marwedel, Combining worst-case timing models, loop unrolling, and static loop analysis for WCET minimization, in *Proceedings of the 22nd Euromicro Conference on Real-Time Systems (ECRTS)*, Dublin, Ireland, July 2009, pp. 35–44

LCFM09.   P. Lokuciejewski, D. Cordes, H. Falk, P. Marwedel, A fast and precise static loop analysis based on abstract interpretation, program slicing and polytope models, in *Proceedings of the International Symposium on Code Generation and Optimization (CGO)*, Seattle, USA, March 2009, pp. 136–146

LFM08.   P. Lokuciejewski, H. Falk, P. Marwedel, WCET-driven cache-based procedure positioning optimizations, in *Proceedings of the 21st Euromicro Conference on Real-Time Systems (ECRTS)*, Prague, Czech Republic, July 2008, pp. 321–330

LFS+07.   P. Lokuciejewski, H. Falk, M. Schwarzer, P. Marwedel, H. Theiling, Influence of procedure cloning on WCET prediction, in *Proceedings of the 5th IEEE/ACM International Conference on Hardware/Software Codesign and System Synthesis (CODES+ISSS)*, Salzburg, Austria, October 2007, pp. 137–142

LFSP07.   P. Lokuciejewski, H. Falk, M. Schwarzer, M. Peter, Tighter WCET estimates by procedure cloning, in *Proceedings of the 7th International Workshop on Worst-Case Execution Time Analysis (WCET)*, Pisa, Italy, July 2007, pp. 27–32

LGM09.   P. Lokuciejewski, F. Gedikli, P. Marwedel, Accelerating WCET-driven optimizations by the invariant path paradigm: a case study of loop unswitching, in *Proceedings of the 12th International Workshop on Software & Compilers for Embedded Systems (SCOPES)*, Nice, France, April 2009, pp. 11–20

LGMM09.   P. Lokuciejewski, F. Gedikli, P. Marwedel, K. Morik, Automatic WCET reduction by machine learning based heuristics for function inlining, in *Proceedings of the 3rd Workshop on Statistical and Machine Learning Approaches to Architecture and Compilation (SMART)*, Paphos, Cyprus, January 2009, pp. 1–15

LKM10.   P. Lokuciejewski, T. Kelter, P. Marwedel, Superblock-based source code optimizations for WCET reduction, in *Proceedings of the 7th IEEE International Conferences on Embedded Software and Systems (ICESS)*, Bradford, UK, June 2010

LPF+10.   P. Lokuciejewski, S. Plazar, H. Falk, P. Marwedel, L. Thiele, Multi-objective exploration of compiler optimizations for real-time systems, in *Proceedings of the 13th IEEE International Symposium on Object/Component/Service-oriented Real-time Distributed Computing (ISORC)*, Carmona, Spain, 2010, pp. 115–122

LSMM10.   P. Lokuciejewski, M. Stolpe, K. Morik, P. Marwedel, Automatic selection of machine learning models for WCET-aware compiler heuristic generation, in *Proceedings of the 4th Workshop on Statistical and Machine Learning Approaches to Architecture and Compilation (SMART)*, Pisa, Italy, January 2010, pp. 3–17

LFK+93.   P.G. Lowney, S.M. Freudenberger, T.J. Karzes et al., The multiflow trace scheduling compiler. J. Supercomput. **7**(1–2), 51–142 (1993)

Lun02.   T. Lundqvist, A WCET analysis method for pipelined microprocessors with cache memories. PhD thesis, Chalmers University of Technology (2002)

LW87.   J.R. Lyle, M.D. Weiser, Automatic program bug location by program slicing, in *Proceedings of International Conference on Computers and Applications*, Peking, China, February 1987, pp. 877–882

Mac02.   P. Machanick, Approaches to addressing the memory wall, Technical report, School of IT and Electrical Engineering, University of Queensland, November 2002

MLC+92.   S.A. Mahlke, D.C. Lin, W.Y. Chen et al., Effective compiler support for predicated execution using the hyperblock. ACM SIGMICRO Newsl. **23**(1–2), 45–54 (1992)

MCG+92.   S.A. Mahlke, W.Y. Chen, J. Gyllenhaal et al., Compiler code transformations for superscalar-based high performance systems, in *Proceedings of the 1992 ACM/IEEE Conference on Supercomputing*, Washington, USA, July 1992, pp. 808–817

MWRG10.   Mälardalen WCET Research Group. WCET Benchmarks, http://www.mrtc.mdh.se/projects/wcet, March 2010

Mar10.   P. Marwedel, *Embedded System Design*, 2nd edn. (Springer, Berlin, 2010)

MM99.   A. McGovern, J.E.B. Moss, Scheduling straight-line code using reinforcement learning and rollouts, in *Proceedings of the Conference on Advances in Neural Information Processing Systems (NIPS)*, Denver, USA, November 1999, pp. 903–909

MMSH01.   G. Memik, W.H. Mangione-Smith, W. Hu, NetBench: a benchmarking suite for network processors, in *Proceedings of the International Conference on Computer-Aided Design (ICCAD)*, San Jose, USA, November 2001, pp. 39–42

MPS94.   A. Mendlson, S.S. Pinter, R. Shtokhamer, Compile time instruction cache optimizations. ACM SIGARCH Comput. Archit. News **22**(1), 44–51 (1994)

Mie08.   I. Mierswa, Non-convex and multi-objective optimization in data mining, PhD thesis, TU Dortmund University, 2008

MWK+06.   I. Mierswa, M. Wurst, R. Klinkenberg et al., YALE: rapid prototyping for complex data mining tasks, in *Proceedings of the 12th ACM SIGKDD International Conference on Knowledge Discovery and Data Mining (KDD)*, Philadelphia, USA, August 2006, pp. 935–940

Min06.   A. Miné, The octagon abstract domain. High.-Order Symb. Comput. **19**(1), 31–100 (2006)

MBQ02.   A. Monsifrot, F. Bodin, R. Quiniou, A machine learning approach to automatic production of compiler heuristics, in *Proceedings of the 10th International Conference* on *Artificial Intelligence: Methodology, Systems, and Applications (AIMSA)*, Varna, Bulgaria, September 2002, pp. 41–50

MPSR95. R. Motwani, K.V. Palem, V. Sarkar, S. Reyen, Combining register allocation and instruction scheduling, Technical report, Stanford University, Stanford, USA, 1995

Mow94. T.C. Mowry, Tolerating latency through software-controlled data prefetching, Technical report, Stanford University, Stanford, USA, 1994

Muc97. S.S. Muchnick, *Advanced Compiler Design and Implementation* (Morgan Kaufmann, San Francisco, 1997)

MG04. S.S. Muchnick, P.B. Gibbons, Efficient instruction scheduling for a pipelined architecture. ACM SIGPLAN Not. **39**(4), 167–174 (2004)

NGE+99. X. Nie, L. Gazsi, F. Engel et al., A new network processor architecture for high-speed communications, in *Proceedings of the IEEE Workshop on Signal Processing Systems (SiPS)*, Taipei, Taiwan, October 1999, pp. 548–557

NNH99. F. Nielson, H.R. Nielson, C. Hankin, *Principles of Program Analysis* (Springer, New York, 1999)

NP93. C. Norris, L.L. Pollock, A schedular-sensitive global register allocator, in *Proceedings of the 1993 ACM/IEEE Conference on Supercomputing*, Portland, USA, November 1993, pp. 804–813

OO84. K.J. Ottenstein, L.M. Ottenstein, The program dependence graph in a software development environment. SIGSOFT Softw. Eng. Not. **9**(3), 177–184 (1984)

PSKL08. A. Pabalkar, A. Shrivastava, A. Kannan, J. Lee, SDRM: simultaneous determination of regions and function-to-region mapping for scratchpad memories. Lect. Not. Comput. Sci. **5374**, 569–582 (2008)

PW86. D.A. Padua, M.J. Wolfe, Advanced compiler optimizations for supercomputers. Commun. ACM **29**(12), 1184–1201 (1986)

PS93. K.V. Palem, B.B. Simons, Scheduling time-critical instructions on RISC machines. ACM Trans. Program. Lang. Syst. (TOPLAS) **15**(4), 632–658 (1993)

PH90. K. Pettis, R.C. Hansen, Profile guided code positioning. ACM SIGPLAN Not. **25**(6), 16–27 (1990)

PLM08. S. Plazar, P. Lokuciejewski, P. Marwedel, A retargetable framework for multi-objective WCET-aware high-level compiler optimizations, in *Proceedings of the 29th IEEE Real-Time Systems Symposium WiP (RTSS)*, Barcelona, Spain, December 2008, pp. 49–52

PLM09. S. Plazar, P. Lokuciejewski, P. Marwedel, WCET-aware software based cache partitioning for multi-task real-time systems, in *Proceedings of the 9th International Workshop on Worst-Case Execution Time Analysis (WCET)*, Dublin, Ireland, June 2009, pp. 78–88

PLM10. S. Plazar, P. Lokuciejewski, P. Marwedel, WCET-driven cache-aware memory content selection, in *Proceedings of the 13th IEEE International Symposium on Object/Component/Service-oriented Real-time Distributed Computing (ISORC)*, Carmona, Spain, 2010, pp. 107–114

PSK08. A. Prantl, M. Schordan, J. Knoop, TuBound—a conceptually new tool for worst-case execution time analysis, in *Proceedings of the 8th International Workshop on Worst-Case Execution Time Analysis (WCET)*, Prague, Czech Republik, July 2008

Pua06. I. Puaut, WCET-centric software-controlled instruction caches for hard real-time systems, in *Proceedings of the 18th Euromicro Conference on Real-Time Systems (ECRTS)*, Dresden, Germany, July 2006, pp. 217–226

PD02. I. Puaut, D. Decotigny, Low-complexity algorithms for static cache locking in multi-tasking hard real-time systems, in *Proceedings of the 23rd IEEE Real-Time Systems Symposium (RTSS)*, Austin, USA, December 2002, pp. 114–123

PP07. I. Puaut, C. Pais, Scratchpad memories vs locked caches in hard real-time systems: a quantitative comparison, in *Proceedings of the Conference on Design, Automation and Test in Europe (DATE)*, Nice, France, March 2007, pp. 1484–1489

RS09. J. Reineke, R. Sen, Sound and Efficient WCET analysis in presence of timing anomalies, in *Proceedings of 9th International Workshop on Worst-Case Execution Time (WCET) Analysis*, Dublin, Ireland, June 2009, pp. 107–117

RWT+06. J. Reineke, B. Wachter, S. Thesing et al., A definition and classification of timing anomalies, in *Proceedings of 6th International Workshop on Worst-Case Execution Time Analysis (WCET)*, Dresden, Germany, July 2006

RR95. T. Reps, G. Rosay, Precise interprocedural chopping, in *Proceedings of the 3rd ACM SIGSOFT Symposium on Foundations of Software Engineering (SIGSOFT)*, Washington, USA, October 1995, pp. 41–52

Ric06. K. Richter, The AUTOSAR timing model—status and challenges, in *Proceedings of the Second International Symposium on Leveraging Applications of Formal Methods, Verification and Validation (ISOLA)*, Paphos, Cyprus, November 2006, pp. 9–10

RTG+07. H. Rong, Z. Tang, R. Govindarajan et al., Single-dimension software pipelining for multidimensional loops. ACM Trans. Archit. Code Optim. **4**(1), 7–51 (2007)

RMC+09. T. Russell, A.M. Malik, M. Chase et al., Learning heuristics for the superblock instruction scheduling problem. IEEE Trans. Knowl. Data Eng. **21**(10), 1489–1502 (2009)

Saa03. A. Saad, Java-based functionality and data management in the automobile. Prototyping at BMW Car IT GmbH, in *Java Spectrum*, March 2003

SEGL06. C. Sandberg, A. Ermedahl, J. Gustafsson, B. Lisper, Faster WCET flow analysis by program slicing. SIGPLAN Not. **41**(7), 103–112 (2006)

Sar01. V. Sarkar, Optimized unrolling of nested loops. Int. J. Parallel Program. **29**(5), 545–581 (2001)

Sch00. H. Schildt, *C/C++ Programmer's Reference* (McGraw-Hill, New York, 2000)

Sch07. D. Schulte, Modeling and transformation of flow facts within a WCET optimizing compiler, Diploma thesis, TU Dortmund University, May 2007

SEG+06. D. Sehlberg, A. Ermedahl, J. Gustafsson et al., Static WCET analysis of real-time task-oriented code in vehicle control systems, in *Proceedings of the 2nd International Symposium on Leveraging Applications of Formal Methods, Verification and Validation (ISOLA)*, Paphos, Cyprus, November 2006, pp. 212–219

Sha89. A.C. Shaw, Reasoning about time in higher-level language software. IEEE Trans. Softw. Eng. **15**(7), 875–889 (1989)

SES00. B. Siegfried, M. Eduard, B. Scholz, Probabilistic procedure cloning for high-performance systems, Technical report, Institute for Software Science, University of Vienna, November 2000

Smi00. M.D. Smith, Overcoming the challenges to feedback-directed optimization, in *Proceedings of the ACM SIGPLAN Workshop on Dynamic and Adaptive Compilation and Optimization (DYNAMO)*, Boston, USA, January 2000, pp. 1–11

SD03. W. So, A. Dean, Procedure cloning and integration for converting parallelism from coarse to fine grain, in *Proceedings of the 7th Workshop on Interaction between Compilers and Computer Architectures (INTERACT)*, Anaheim, USA, February 2003, pp. 27–36

SK04. L. Song, K. Kavi, What can we gain by unfolding loops? SIGPLAN Not. **39**(2), 26–33 (2004)

SLPH+05. J. Souyris, E. Le Pavec, G. Himbert et al., Computing the worst case execution time of an avionics program by abstract interpretation, in *Proceedings of the 5th International Workshop on Worst-Case Execution Time Analysis (WCET)*, Palma de Mallorca, Spain, June 2005, pp. 21–49

SS07. Y.N. Srikant, P. Shankar, *The Compiler Design Handbook: Optimizations and Machine Code Generation*, 2nd edn. (CRC Press, Boca Raton, 2007)

Sta88. J.A. Stankovic, Misconceptions about real-time computing: a serious problem for next-generation systems. Computer **21**(10), 10–19 (1988)

SEE01. F. Stappert, A. Ermedahl, J. Engblom, Efficient longest executable path search for programs with complex flows and pipeline effects, in *Proceedings of the 2001 International Conference on Compilers, Architecture, and Synthesis for Embedded Systems (CASES), Atlanta, USA, November 2001, pp. 132–140

SW+10. S. Steinke, L. Wehmeyer et al., The encc Energy aware C compiler homepage, http://ls12-www.cs.tu-dortmund.de/research/activities/encc, March 2010

SA05. M. Stephenson, S. Amarasinghe, Predicting unroll factors using supervised classification, in Proceedings of the International Symposium on Code Generation and Optimization (CGO), San Jose, USA, March 2005, pp. 123–134

SAMO03. M. Stephenson, S. Amarasinghe, M. Martin, U.M. O'Reilly, Meta optimization: improving compiler heuristics with machine learning. SIGPLAN Not. 38(5), 77–90 (2003)

SDJ84. B. Su, S. Ding, L. Jin, An improvement of trace scheduling for global microcode compaction. ACM SIGMICRO Newsl. 15(4), 78–85 (1984)

SM08. V. Suhendra, T. Mitra, Exploring locking & partitioning for predictable shared caches on multi-cores, in Proceedings of the 45th annual Design Automation Conference (DAC), Anaheim, California, June 2008, pp. 300–303

SRM08. V. Suhendra, A. Roychoudhury, T. Mitra, Scratchpad allocation for concurrent embedded software, in Proceedings of the 6th IEEE/ACM International Conference on Hardware/Software Codesign and System Synthesis (CODES+ISSS), Atlanta, USA, October 2008, pp. 37–42

SMR+05. V. Suhendra, T. Mitra, A. Roychoudhury et al., WCET centric data allocation to scratchpad memory, in Proceedings of the 26th IEEE International Real-Time Systems Symposium (RTSS), Miami, USA, December 2005, pp. 223–232

Sym10. Symtavision GmbH, Scheduling Analysis for ECUs, Buses and Networks, http://www.symtavision.com/symtas.html, March 2010

Inf08a. Tc1796 32-bit single-chip microcontroller tricore—data sheet. Infineon Technologies AG, Document Revision 2008-04 (2008)

The00. H. Theiling, Extracting safe and precise control flow from binaries, in Proceedings of the Seventh International Conference on Real-Time Systems and Applications (RTCSA), Cheju Island, South Korea, December 2000, pp. 23–30

The02. H. Theiling, Control flow graphs for real-time systems analysis, PhD thesis, Saarland University, 2002

The04. S. Thesing, Safe and precise WCET determinations by abstract interpretation of pipeline models, PhD thesis, Saarland University, 2004

TCG+02. L. Thiele, S. Chakraborty, M. Gries et al., A framework for evaluating design tradeoffs in packet processing architectures, in Proceedings of the 39th Design Automation Conference (DAC), New Orleans, USA, June 2002, pp. 880–885

TY97. H. Tomiyama, H. Yasuura, Code placement techniques for cache miss rate reduction. ACM Trans. Des. Automat. Electron. Syst. 2(4), 410–429 (1997)

Ton99. P. Tonella, Effects of different flow insensitive points-to analyses on DEF/USE sets, in Proceedings of the Third European Conference on Software Maintenance and Reengineering (CSMR), Amsterdam, The Netherlands, March 1999, pp. 62–69

Inf04. Tricore 1 pipeline behaviour & instruction execution timing. Infineon Technologies AG, Document Revision 2004-06 (2004)

Inf08b. TriCore 1 32-bit unified processor core v1.3 architecture—architecture manual. Infineon Technologies AG, Document Revision 2008-01 (2008)

UTD10. UTDSP Benchmark Suite. http://www.eecg.toronto.edu/~corinna/DSP/infrastructure/UTDSP.html, March 2010

Vah99. F. Vahid, Procedure cloning: a transformation for improved system-level functional partitioning. ACM Trans. Des. Automat. Electron. Syst. 4(1), 70–96 (1999)

Vah02. F. Vahid, Embedded System Design: A Unified Hardware/Software Introduction (Wiley, New York, 2002)

Vah08. F. Vahid, Timing is everything—embedded systems demand teaching of structured time-oriented programming, in Proceedings of the Workshop on Embedded Systems Education (WESE), Atlanta, USA, October 2008, pp. 1–9

VTS+07. K. Vaswani, M.J. Thazhuthaveetil, Y. Srikant et al., Microarchitecture sensitive empirical models for compiler optimizations, in *Proceedings of the International Symposium on Code Generation and Optimization (CGO)*, San Jose, USA, March 2007, pp. 131–143

VRF+08. H. Venturini, F. Riss, J.C. Fernandez et al., A fully-non-transparent approach to the code location problem, in *Proceedings of the 11th International Workshop on Software & Compilers for Embedded Systems (SCOPES)*, Munich, Germany, March 2008, pp. 61–68

VLX03. X. Vera, B. Lisper, J. Xue, Data cache locking for higher program predictability, in *Proceedings of the ACM SIGMETRICS International Conference on Measurement and Modeling of Computer Systems (SIGMETRICS)*, San Diego, USA, July 2003, pp. 272–282

Ver09. S. Verdoolaege, Barvinok, http://www.kotnet.org/~skimo/barvinok, September 2009

VSB+04. S. Verdoolaege, R. Seghir, K. Beyls et al., Analytical computation of Ehrhart polynomials: enabling more compiler analyses and optimizations, in *Proceedings of the 2004 International Conference on Compilers, Architecture, and Synthesis for Embedded Systems (CASES)*, Washington, USA, September 2004, pp. 248–258

VM07. M. Verma, P. Marwedel, *Advanced Memory Optimization Techniques for Low-Power Embedded Processors* (Springer, Berlin, 2007)

WHSB92. N.J. Warter, G.E. Haab, K. Subramanian, J.W. Bockhaus, Enhanced modulo scheduling for loops with conditional branches. ACM SIGMICRO Newsl. **23**(1–2), 170–179 (1992)

WM06. L. Wehmeyer, P. Marwedel, *Fast, Efficient and Predictable Memory Accesses: Optimization Algorithms for Memory Architecture Aware Compilation* (Springer, New York, 2006)

Wei99. M.D. Weiser, The computer for the 21st century. ACM SIGMOBILE Mobile Comput. Commun. Rev. **3**(3), 3–11 (1999)

Wei79. M.D. Weiser, Program slices: formal, psychological, and practical investigations of an automatic program abstraction method, PhD thesis, University of Michigan, Ann Arbor, USA, 1979

WS90. D. Whitfield, M.L. Soffa, An approach to ordering optimizing transformations. ACM SIGPLAN Not. **25**(3), 137–146 (1990)

Wol02. F. Wolf, *Behavioral Intervals in Embedded Software: Timing and Power Analysis of Embedded Real-Time Software Processes* (Kluwer Academic, Norwell, 2002)

WM95. W.A. Wulf, S.A. McKee, Hitting the memory wall: implications of the obvious. ACM SIGARCH Comput. Archit. News **23**(1), 20–24 (1995)

WM07. M. Wurst, K. Morik, Distributed feature extraction in a P2P setting—a case study. Future Gener. Comput. Syst. **23**(1), 69–75 (2007). Special Issue on Data Mining

ZCS03. M. Zhao, B. Childers, M.L. Soffa, Predicting the impact of optimizations for embedded systems, in *Proceedings of the ACM SIGPLAN Conference on Languages, Compilers, and Tools for Embedded Systems (LCTES)*, San Diego, USA, June 2003, pp. 1–11

ZWHM04. W. Zhao, D. Whalley, C. Healy, F. Mueller, WCET code positioning, in *Proceedings of the 25th IEEE International Real-Time Systems Symposium (RTSS)*, Lisbon, Portugal, December 2004, pp. 81–91

ZWH+05. W. Zhao, D. Whalley, C. Healy et al., Improving WCET by applying a WC code-positioning optimization. ACM Trans. Archit. Code Optim. **2**(4), 335–365 (2005)

ZKW+05. W. Zhao, W. Kreahling, D. Whalley et al., Improving WCET by optimizing worst-case paths, in *Proceedings of the 11th IEEE Real Time on Embedded Technology and Applications Symposium (RTAS)*, San Francisco, USA, March 2005, pp. 138–147

ZKW+04. W. Zhao, P. Kulkarni, D. Whalley et al., Tuning the WCET of embedded applications, in *Proceedings of the 10th IEEE Real-Time and Embedded Technology and Applications Symposium (RTAS)*, Toronto, Canada, May 2004, pp. 472–481

ZK04.   E. Zitzler, S. Künzli, Indicator-based selection in multiobjective search, in *Proceedings of the 9th International Conference on Parallel Problem Solving from Nature (PPSN)*, Birmingham, UK, September 2004, pp. 832–842

ZT99.   E. Zitzler, L. Thiele, Multiobjective evolutionary algorithms: a comparative case study and the strength Pareto approach. IEEE Trans. Evol. Comput. 3(4), 257–271 (1999)

ZLT01.   E. Zitzler, M. Laumanns, L. Thiele, SPEA2: improving the strength Pareto evolutionary algorithm for multiobjective optimization, in *Proceedings of the Conference on Evolutionary Methods for Design, Optimisation and Control with Application to Industrial Problems (EUROGEN)*, Athens, Greece, September 2001, pp. 95–100

ZVS+94. V. Zivojnović, J. Martínez Velarde, C. Schläger et al., DSPstone: a DSP-oriented benchmarking methodology, in *Proceedings of the International Conference on Signal Processing and Technology (ICSPAT)*, Dallas, USA, January 1994, pp. 715–720

Index

P. Lokuciejewski, P. Marwedel, *Worst-Case Execution Time Aware*
Compilation Techniques for Real-Time Systems, Embedded Systems,
DOI 10.1007/978-90-481-9929-7, © Springer Science+Business Media B.V. 2011